THE JACK-UP DRILLING PLATFORM
Design, Construction and Operation

Papers presented at the Second International Conference on The Jack-Up Drilling Platform: Design, Construction and Operation, held at City University, London, UK, 26–27 September 1989.

Organised by:

Ocean Engineering Research Centre, Department of Civil Engineering, City University, London, UK.

Sponsored by:

The Institution of Mechanical Engineers
The Royal Institute of Naval Architects
The Steel Construction Institute
Noble Denton & Associates

THE JACK-UP DRILLING PLATFORM
Design, Construction and Operation

Edited by

L. F. BOSWELL and C. A. D'MELLO

Ocean Engineering Research Centre
Department of Civil Engineering
City University
London, UK

Reprinted from

Marine Structures

Vol. 2 Nos 3–5

ELSEVIER APPLIED SCIENCE
London and New York

ELSEVIER SCIENCE PUBLISHERS LTD
Crown House, Linton Road, Barking, Essex IG11 8JU, England

Sole Distributor in the USA and Canada
ELSEVIER SCIENCE PUBLISHING CO., INC.
52 Vanderbilt Avenue, New York, NY 10017, USA

WITH 39 TABLES AND 141 ILLUSTRATIONS

© 1990 ELSEVIER SCIENCE PUBLISHERS LTD

British Library Cataloguing in Publication Data

The Jack-up drilling platform: design, construction and
operation.
1. Mobile offshore drilling rigs. Design
I. Boswell, L. F. II. D'Mello, C. A. III. Marine
structures
627.98

ISBN 1-85166-490-4

Library of Congress CIP data applied for

Phototypesetting by Tech-Set, Gateshead, Tyne & Wear.
Printed in Great Britain by Galliard (Printers) Ltd, Great Yarmouth.

Preface

Jack-Up drilling platforms are used for the exploration and operation of offshore oil and gas fields, as well as for the servicing of fixed structures. Originally designed for use in shallow waters, they are being used increasingly in deeper conditions. They have the major advantage of being re-usable so helping marginal field development. Jack-Up drilling platforms contribute to a significant part of offshore engineering activity around the world.

The application of the platforms is continuously being extended towards deeper waters and harsher environments. Designs have been developed for operation in areas with maximum wave heights of 30 m and water depths well beyond 100 m.

The first international conference relating to the Jack-Up platform in 1985 was a great technical success and the papers which were presented represented a landmark in the presentation of relevant technical material.

Much has happened within the oil and gas industry since 1985, but the realisation to maintain an effort to understand and agree upon the performance of Jack-Ups has always remained. It seemed quite natural, therefore, to organise another conference in which more recent and relevant material would be presented.

A number of areas are of considerable interest to the Jack-Up industry and these include the calculation of hydrodynamic coefficients, the consideration of safety factors, the effect of marine growth, current profiles, second order effects, the weldability of high strength steels and collision effects.

The second international conference considers many aspects of these areas and a particular attempt has been made to present associated material. The papers presented at the conference have been written by practising professionals who are recognised experts from the UK, Europe and the USA.

This volume, therefore, represents a further stage in the presentation of material to improve our understanding of the behaviour of Jack-Up platforms.

L. F. Boswell
December 1989

Contents

Contents

Problems with Certification and Legislation

Richard E. Haas

Rowan Companies, Inc., 5051 Westheimer, Houston, Texas 77056, USA

ABSTRACT

This paper discusses the vast growth of requirements directed toward mobile offshore drilling units attempting to be internationally certified. As a simplistic example, the regulations regarding heliports are examined with the pitfalls and benefits of an international standard discussed from the viewpoint of owner and operator. Recommendations contained in an International Maritime Organization resolution are suggested for use by the various regulatory bodies when dealing with new or amended rules. It is suggested that a more universal standard would evolve if the guidance of the resolution were heeded.

Key words: compelling need, harmonization, international standards.

On 15 November 1979, the Assembly of the then Inter-Governmental Maritime Consultative Organization (IMCO)[†] adopted Resolution A.414 (XI), Code for the Construction and Equipment of Mobile Offshore Drilling Units (MODU Code). Five days later, the Assembly adopted Resolution A.500 (XII) concerning the objectives of the organization in the 1980s. Resolution A.500 recommends that proposals for new conventions or amendments to existing conventions be entertained only 'on the basis of clear and well-documented demonstration of compelling need, taking into account the undesirability of modifying conventions not yet in force or of amending existing conventions unless

---

[†] The Inter-Governmental Maritime Consultative Organization changed its name to the International Maritime Organization (IMO) in 1982.

such latter instruments have been in force for a reasonable period of time and experience has been gained of their operation, and having regard to the costs to the maritime industry and the burden on the legislative and administrative resources of Member States'.

The MODU Code was developed to provide an international standard for new construction to facilitate international movement and operation of these units. It was never intended that the Code would prohibit the use of an existing unit simply because its design, construction and equipment did not conform to the requirements of the Code. Governments were invited to 'give effect to the Code not later than 31 December 1981.' After all these eloquent phrases, what has happened is that only a few of the lesser-developed nations of the world have adopted the Code as their national regulations and some of the other lesser-developed nations have required units to possess a valid IMO MODU Code certificate in order to drill on their continental shelves. The more advanced countries have developed their own regulations for their MODUs and have required compliance by foreign-flag units operating in their waters. Compliance with the IMO MODU Code in these cases means virtually nothing to the coastal state.

In 1982, Norway submitted a paper (DE 25/9) to the IMO Ship Design and Equipment (D & E) Subcommittee calling for a review of the MODU Code. The justification for this was the tragic accident to the *Alexander L. Kielland*, an accommodation unit which, although originally constructed as a drilling unit, never operated as such. The *Kielland* construction contract was signed in 1973 and the completed platform delivered in July 1976. This was over 3 years prior to the adoption of the MODU Code by the IMO and 5 years prior to the invited effective date for individual nations. The paper also cited the loss of another semi-submersible unit, *Ocean Ranger*, in February 1982 as a need for reconsidering the standards even though it too had been built in 1976 and neither the flagging nation (USA) nor the coastal state (Canada) had completed even preliminary investigations of this tragic accident.

This initial paper addressed sections in 8 of the 14 chapters of the IMO MODU Code as in need of revision. The International Association of Drilling Contractors (IADC) studied the submission as well as the official report covering the *Kielland* accident and could not determine the compelling need to amend the Code based on the facts presented. The presentation by the IADC (DE 26/8) fell on deaf ears and by the January 1985 session, the entire Code was opened for discussion with papers submitted on all Chapters except those covering Lifting devices and Helicopter facilities. The Resolution A.500 recommendations regarding clear and well-documented demonstrations of compelling needs were lost in the rush to revise the Code.

'Compelling need' was replaced by 'demonstrated need' which was in turn replaced by 'whatever seemed desirable'. Whenever the aspect of 'having regard to costs to the maritime industry' was raised, the response was that money is of no object when it comes to safety. While the MODU Code was undergoing the revision, another item appeared on the IMO Ship Design and Equipment Subcommittee agenda which was separate and distinct, 'Helicopter Facilities Offshore (IMO/ICAO Agreement).' Treated as another topic on the work agenda of a general nature, it did result in an agreement between IMO and the International Civil Aviation Organization (ICAO) on 'helicopter operations to and from ships and other marine vehicles.' In March 1985, the Memorandum of Understanding... 'Concerning Cooperation in Respect of Safety of Aircraft Operations to and From Ships and Other Marine Vehicles...' was signed. The Agreement stipulated that all matters which are directly connected with the design, construction and equipment of ships and other marine vehicles and their operation should be regarded as falling primarily within the field of responsibility of IMO. With all the other MODU Code chapters being reviewed and no papers submitted on MODU Helicopter Facilities, the topic was dismissed as inconsequential by the drilling contractor representatives, a move which has come back to haunt them.

The deliberations ongoing within the IMO did not preclude other entities and countries from establishing their own standards. Concurrently a major oil company launched a program purporting to upgrade marine safety standards and practices on MODUs. Based on this initiative, the American Petroleum Institute (API) formed a subcommittee to identify, monitor, and coordinate marine safety task actions. The American Bureau of Shipping, as well as other classification societies, increased their rulemaking efforts with regard to the MODU safety aspects. The United States Coast Guard issued regulations and proposed rules for comment including their entire subchapter of Federal Regulations regarding MODUs. The United Kingdom Department of Energy completed a draft fourth edition of their guidance on design, construction and certification of offshore installations. The Canadian government promulgated standards covering design and construction of MODUs based on the IMO Code and studies regarding the loss of the *Ocean Ranger*. The North Sea countries were all issuing their own rules which due to the variances between nations led to a move to develop a set of North West European Standards. This harmonization effort could not resolve the differences and no uniform set of rules to facilitate movement even within this geographic area could be drafted and agreed upon.

Each entity has not only published their own guidelines, many also presented their rules as input for consideration in the MODU Code review. Again, the well-documented compelling need criteria had long since been discarded in the discussions within the IMO Working Groups. Negotiations and concessions were made at the IMO sessions recognizing that the effort was revising a code which is advisory in nature and not mandatory. However, after each session, decisions made regarding the Code have been reflected in other organizations' rules or regulations as compulsory requirements. Although the new version of the IMO Code is intended to be applicable only to new construction, as was the original Code, when portions are reproduced as national rules in many instances the application becomes universal with no 'grand-fathering' being provided to existing units. The process from which the new MODU Code evolved has been a result of various inputs on a continuing basis over the past six years. The target completion date for the IMO Ship Design and Equipment Subcommittee was their 32nd session set for December 1988. The report of the 31st session stated:

'The Sub-committee noted that the work on the revision of the MODU Code, with the exception of chapters 3 and 9, has been finalized and any comments that may be submitted to the next session should only refine and harmonize the wording in the chapters and no new major issues should be presented.' DE 31/16 para. 4.45. dated 23 March 1988.

Submissions regarding the revision to the MODU code had totalled 33 through the 28th session of the Ship Design and Equipment Sub-committee (DE 28) held in January 1985. For the next session, DE 29 held in May 1986 the total new documents to be considered was 38. DE 30 in June 1987 was inundated by 55 more papers for review and possible incorporation into the revised Code. In March 1988, the 31st session had only 27 new documents to digest prior to closing all chapters other than 3 and 9. By this time, every chapter, except Chapter 13 — Helicopter Facilities — had been addressed and/or readdressed in submissions from the member nations, observer organizations and the Secretariat. Then on July 4, 1988, the first paper (DE32/5) published for consideration at the December 1988 session was a 'Note by the Secretariat' completely rewriting the chapter on 'Helicopter Facilities'. Subsequently 15 more papers were submitted for review in conjunction with revising the MODU Code including two which objected to the Secretariat paper. The majority of these final submissions addressed specifically Chapters 3 and

9 with some other minor areas of concern being clarified in the remainder. With the very limited time available at the December 1988 Ship Design and Equipment Subcommittee meeting to complete the revision of the Code, agreement with the rewritten Chapter 13 could not be reached by the working Group. The Subcommittee agreed not to make any changes to the chapter, but agreed to add a footnote to the chapter title referencing applicable standards of ICAO and the Memorandum of Understanding between IMO and ICAO.

The presentation of the rewritten chapter on helicopter facilities is symptomatic of the problems encountered in legislation and certification. With over 160 papers presented directly addressing the review of the IMO MODU Code, the drilling contractor representatives failed to grasp the possible consequences of the ICAO/IMO agreement agenda item. In January 1985, the DE Subcommittee did request that ICAO comment on the existing IMO text regarding helicopter facilities. At this same meeting it was decided that a small IMO correspondence group would be established to assist ICAO with respect to marine matters and that the IMO would have a member of the Secretariat attend relevant ICAO meetings. The May 1986 meeting report notes that IMO comments on fire protection matters, visual aids and helicopter operations in the revised annex to the ICAO were up-to-date and no further action was needed. With regard to Chapter 13 of the MODU Code, the Subcommittee agreed to take no further action until the revised Annex 14 was *finally adopted* by ICAO. The June 1987 meeting report regarding helicopter facilities offshore was essentially the same as the one the previous year.

At the DE 31 session, March 1988, the meeting report reveals with respect to helicopter facilities offshore, there were no submissions to the session and no formal correspondence between ICAO and IMO either requesting information or comments on matters of common interest. The Subcommittee agreed to delete the item from future agenda but to keep it in the work program for review and reactivation as needed. The Secretariat was to monitor events and keep the Subcommittee informed. It was with this background that the revised Chapter 13 was submitted at the final session for adoption into the revised MODU Code and, as of this point in time, the revised Annexs 6 and 14 to the ICAO Convention have still not been adopted.

An international standard accepted universally would be very desirable from the viewpoint of both the drilling unit designer and owner. During the design phase of a MODU, the helicopter requirements are but one aspect of the entire package to be researched and incorporated into the construction. To provide a realistic example, this paper will address a hostile-environment unit as the MODU to be designed and the Sikorsky S-61 helicopter as the design helicopter.

The initial aspect to be investigated is the size of the heliport which is dictated by the largest aircraft intended to use the facility, the 'design helicopter'. For the North Atlantic and North Sea areas, the required size varies from the rotor diameter (18·90 m) to the overall length (22·3 m) to 1·33 times the rotor diameter (25·14 m)depending upon which coastal state regulations are chosen. The obstacle-free approach/departure sector prescribed in the same areas is either 180 or 210 degrees, again dependent upon coastal state chosen. Not only do the design load factors vary in the structural design but the analytical approach required is working stress or limit-state again dependent upon the country involved. The designed structure is mandated by the most stringent criteria for each category in the desired operational area. Once designed, constructed and placed in operation, the MODU is continually reassessed based on, at times, new criteria. This could be instigated by coastal state regulation or possibly oil company policy. Generally the heliport structure, given reasonable research prior to construction, is considered adequate for size and strength however the accessories and marking are continually subject to change.

For instance, to relocate a drilling unit from the United Kingdom sector of the North Sea to the Netherlands sector requires repainting of the heliport. Items which differ include the width of the perimeter marking (0·3 m *vs* 0·4 m), diameter of aiming circle (14 m *vs* 13·1 m), position of aiming circle (9 m from outboard edge *vs* concentric to landing area) and size of the prescribed 'H' (4 m *vs* 3 m). Other items which vary are the number and placement of the diameter markings, maximum permitted height of landing lights and the color of the perimeter lights. These changes are required although the MODU may have moved less than 5 km and is being serviced by the same type aircraft. If we expand the area of comparison to include other North Sea countries or other North Atlantic areas, the differences become even greater. This is only one aspect of the MODU which must be altered to shift from one sector to another in the North Sea, yet it is one which could be made uniform without much effort. Such a harmonization would certainly be applauded by the MODU owners, although the paint contractors may protest.

Having said that harmonization would be welcomed, why would the revised chapter to the IMO MODU Code be such a point of contention? For the North Sea region, the Secretariat paper would be acceptable. However, the paper alters the construction of the original chapter by eliminating the criteria for the benign weather areas of the world. The original MODU Code provides a set of general criteria for most areas and a second set for those areas 'where adverse climatic conditions are prevalent, as in the North Sea.' The elimination of this 'two-tier' proviso would increase the required heliport diameter from 18·90 m to 22·3 m throughout the general areas of the world. In terms of weight and cost, the increases are approximately 40 tons and $125 000. These increases are based strictly on the change of criteria from rotor diameter to overall length on the representative S-61 helicopter and do not include changes in design strength calculations.

During the IMO working group meeting in December 1988, two countries brought their ICAO representatives to the sessions involving the helicopter facilities. During these meetings it was possible to pass information concerning MODU operations which the ICAO representatives were completely unaware of and that changed their viewpoints regarding certain facets of the proposed regulations. A correspondence group consisting of nine members or observer groups was formed to try to resolve the variances between the original MODU Code and the paper by the Secretariat. Besides the formation of this correspondence group, it was decided that a joint IMO/ICAO meeting was to be conducted in Montreal in mid-1989 to cover various aspects regarding the drilling operations in order to attempt to rationalize the differences. Although the Subcommittee had requested the Secretariat to arrange the meeting, and one was tentatively set for early July, it was subsequently cancelled by the IMO staff. The correspondence group is still attempting to reach a consensus even though no demonstration of compelling need has ever been produced. The only reasoning behind the complete rewrite has been to attempt to harmonize the IMO MODU Code with the ICAO proposed documents.

While the IMO has been revising the MODU Code and ICAO amending its conventions, the Helicopter Safety Advisory Committee (HSAC) has produced and published under the sponsorship of the Louisiana Department of Transportation and Development a second revision to their 'Offshore Heliport Design Guide'. The HSAC is a voluntary safety organization consisting of oil companies, helicopter operators and government agencies with the dedicated aim of improving safety offshore. The guide presents criteria which are specifically directed at operations in the Gulf of Mexico but would possibly be applicable to the general areas of the world. The American Petroleum Institute (API) also produced in 1986 a recommended practice for 'Planning, Designing and Constructing Heliports for Fixed Offshore Platforms'. The American Bureau of Shipping (ABS) and other classification societies address helicopter decks in their rules for mobile offshore drilling units. In May 1989, the International Chamber of Shipping (ICS) published the third edition of their 'Guide to Helicopter/ Ship Operations'. The United States Federal Aviation Administration (FAA) published a revised heliport design advisory circular in 1988 for public distribution.

These are the recently issued publications and the list is not presumed to be all-encompassing yet it demonstrates that many varied organizations are addressing heliports. It is ironic that the only one of the group mentioned above in attendance at the ICAO deliberations regarding heliports on MODUs was the FAA who considered them as private facilities and hence not in the purview of the ICAO convention. With this determination, the FAA opted to attend the meetings as observers rather than members.

In the course of the review of the IMO MODU Code, the drilling contractor representatives relied upon the IMO reports regarding status of ICAO progress in helicopter facilities. The work performed by API and HSAC was neither known of nor monitored by the International Association of Drilling Contractors (IADC). There are drilling representatives on the ABS special committee which review the changes to the MODU rules for classification; however, the only change from 1980 to 1988 regarding heliport design was the addition of a paragraph requiring the taking into account the entire unit motion. The ICS publication, unknown to the drilling industry prior to the latest edition, addresses vessels of all types including gas and oil carriers. With independent documents being produced on a regular basis and although not addressing MODUs directly, many can and do present concepts that will later evolve into the regulations regarding the drilling units.

At the December 1988 IMO meeting, the ICAO representatives stated that several countries of the world have already adopted the proposed ICAO revisions as their national regulations. Two of the countries mentioned are of the North Sea community while at least two more in the moderate areas of the world were also indicated to have accepted the international standards. These same requirements are being inserted by some oil companies into the bid packages soliciting drilling units in various areas of the world without regard to climatic conditions. At least one bid solicitation requires the new ICAO criteria in a moderate area of the world and states that the design aircraft must be the S-61 and further that the aircraft serving the facility will be the smaller Aerospatiale Puma 330J helicopter. This dictates that the diameter of the landing area must be at least 22·3 m whereas the servicing aircraft has an overall length of 18·2 m (less than the rotor diameter of the design helicopter). If the original IMO MODU Code for the general areas applied, the required diameter would be only 15·0 m which is the rotor diameter of the Puma. The bid request states that no exceptions would be granted regarding the helicopter facility requirements. It is this type of growth of unjustified mandates that should be discussed and analyzed by some conference with a view toward slowing the avalanche.

Helicopter Facilities is but one chapter in fourteen of the IMO MODU Code and is certainly not the most comprehensive or intricate. However, it does serve as an excellent illustration of some problems entailed with international movement of drilling units. The code is, by definition, for the construction and equipment of Mobile Offshore Drilling Units and specifically states that drilling operations are subject to the requirements of the coastal states. These coastal state rules have evolved to cover far more than merely drilling operations while the Code is not even recognized by the loudest proponents for revision.

In recent years, the customers of the drilling contractors, the oil companies, have instituted their own stipulations and criteria which in many instances far exceed governmental requirements. The new revised IMO MODU Code has not yet been officially sanctioned by its highest body, the Assembly, and already one of the most active participating countries has approached the International Standards Organization (ISO) to develop a set of standards for mobile offshore drilling units so as to permit international movement with a minimum of problems and modifications. The International Electrotechnical Commission (IEC) has already initiated a project to develop electrical regulations for the MODUs. Presently in conjunction with this conference is a meeting of a joint industry study on jackup assessment standards. This study is off and running on the development of a technical guideline for assessing jackup locations worldwide. What started as a local concern is now evolving to worldwide policy without ever producing a demonstrable, let alone compelling, need. The list of studies and projects, when considering national as well as international, is seemingly endless.

The drilling contractor companies have been trying to financially survive for the past several years with most organizations incurring substantial losses. Contractors have ceased operations, reduced staffs, combined, reorganized, and been absorbed by financial institutions while these regulatory initiatives have been evolving. In the survival mode, the drilling contractors have had to severely limit participation in these ventures due to lack of resources. In many instances, one person within an organization is all that is available to try to respond to each of the proposals within their scope of operation. With the incessant stream of outside demands, only the most urgent requests can be accommodated and the priority is generally mandated according to areas of possible rig employment. Many man-years of effort from the contractors were expended during the revision of the IMO MODU Code since it had the greatest visibility. Consequently active involvement in other projects was limited due to constraints within the companies, lack of knowledge of the undertaking, or lack of invitation to participate.

The MODU Code was created due to the uniqueness of the vessels, with the expressed purpose of facilitating international movement and operation while providing an equivalent level of safety with conventional ships. The Code does not prohibit the use of an existing unit because its design, construction and/or equipment fail to meet that prescribed by the document. The authors were well-intentioned; however, as a viable, useful document, the IMO MODU Code certificate has failed to gather much acceptance. The concept still has merit and there are many drilling contractors who would gladly attain a universally-recognized document, if one were available. Harmonization would invariably lessen the costs of certification, as shown in the case of heliports. However, the criteria can not be mandated strictly by the blind adoption of the most stringent regulations. Consideration should be given to the environment and conditions affecting the drilling and support operations when attempting to develop an all-inclusive set of requirements. The forum for such an undertaking would have to be an acknowledged and accepted authority prior to commencement of the project. Such a concept is utopian as the chance of agreement universally is virtually zero. The most that can be reasonably expected would be that the Codes are accepted for what they are and given acceptance by other than a few nations scattered throughout the world. Future standard drafting efforts, whether by governmental and/or industrial bodies, would be well advised to heed the concepts expressed in IMO Resolution A.500 (XII) paragraphs 3 and 4.

3. Recommends that the Council and Committees entertain proposals for new conventions or amendments to existing conventions only on the basis of clear and well-documented demonstration of compelling need, taking into account the undesirability of modifying conventions not yet in force or of amending existing conventions unless such latter instruments have been in force for a reasonable period of time and experience has been gained of their operation, and having regard to the costs to the maritime industry and the burden on the legislative and administrative resources of Member States.

4. Recommends that the committees carrying out their functions on the basis of the principle that provisions of new conventions or of amendments to existing conventions relating to the structure of

ships should apply only to ships built after the entry into force of the instrument or instruments in question and that other provisions should not apply to ships built before the entry into force of the instrument or instruments in question unless there is a compelling need and the costs and benefits of the measures have been fully considered.

Given this approach to the initiation or alteration of standards, the chances of harmonization between authorities would be greatly enhanced. Abiding by these simple recommendations would not inhibit essential progress but would prevent the instigation of demands without any apparent underlying reasoning. This doctrine would eliminate many of the problems of legislation and certification presently encountered by the international drilling contractors and allow them the opportunity to concentrate on true safety issues rather than sorting through a myriad of different and often conflicting demands.

Jack-Up Conversion for Production

L. C. Scot Kobus, Robert W. Fogal

BGMB (USA) Inc., PO Box 671845, Houston, Texas, USA

&

Egidio Sacchi

BGMB S.P.A., Milano, Italy

ABSTRACT

For the production development of marginal offshore oil fields or to achieve early production in offshore fields scheduled for full production facilities, a jack-up drilling rig can be converted to operate as a mobile production unit. This paper presents a summary discussion of the main aspects to be considered in the design, conversion, installation and operation of a jack-up production unit.

Key words: production, jack-up rig, conversion, early production, marginal production, extended well testing.

1 INTRODUCTION

Production development of offshore oil fields has been traditionally accomplished by means of a rigid structure that is installed in a fixed position resting on the seafloor at an offshore field site. These platforms typically have lower-section structures that are constructed of either steel or concrete, upon which are constructed steel deck structures. These upper structures provide the platform decks on which the production process equipment and operating-personnel accommodations are installed.

In the early 1970s, when the price of crude oil increased rapidly, oil companies intensified their efforts to develop methods by which to exploit small offshore oil fields that could not economically justify the installation of any type of conventional fixed-platform facilities based on the results of an exploratory drilling program. These small fields are designated as marginal fields.[1]

There is typically a 4–6 year time lag between the confirmation of a field's production capabilities, based on the evaluation of the results of an exploratory drilling program, and the initiation of production from a fixed-platform facility. The oil companies also sought methods by which they could accelerate production in offshore fields that had been selected for the installation of conventional production facilities.

One of the developed methods involved converting semi-submersible and jack-up mobile offshore drilling rigs for production operations. Another method involved modifying tankers and barges. In all cases, the conversion or modification involved the installation of production equipment that would normally be placed on a fixed platform.

1.1 Semi-submersible production units

At offshore sites where the water depth exceeds 100 m and/or the environmental conditions are categorized as being severe in nature, converted semi-submersible drilling rigs have been used in production operations. These units have performed successfully, notably in the severe environment of the North Sea and at deep water sites of offshore Brazil.[2] (Fig. 1).

Based on the success of these converted semi-submersibles for production operations, offshore designers have developed semi-submersible units specifically intended to be built for production service. Some of these units are designed to have both drilling and production capabilities. To date, only one such unit has been constructed and placed in service.

1.2 Tanker or barge production units

There have been numerous applications of the use of converted tankers or barges for marginal field or early production service[3] (Fig. 2). Converted tankers or barges have the favorable feature of possessing suitable internal compartmentation and carrying capacity to enable onboard storage of processed oil, whereas converted semi-submersibles

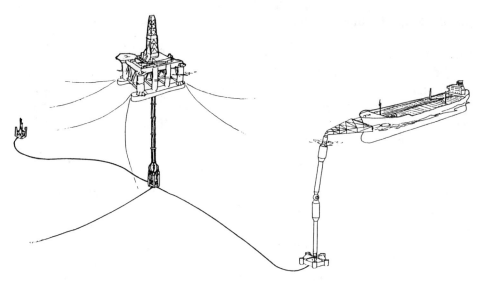

Fig. 1. Semi-submersible production unit.

and jack-up drilling rigs do not. To date, no existing drillships have been converted for production operations.

However, since tankers and barges do not provide a motion-stable platform on which to conduct production operations in moderate or severe environmental conditions, their use has been confined to offshore regions with mild environmental conditions.

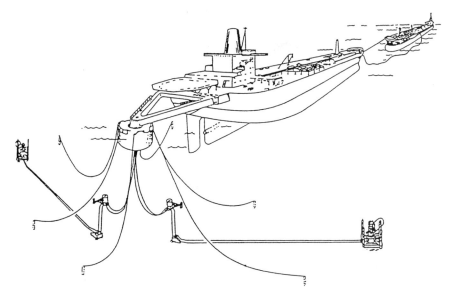

Fig. 2. Tanker production unit.

1.3 Jack-up production units

Since the most extensive demand for marginal field or early production has been in the severe environmental areas of the North Sea and the deep water regions of offshore Brazil, converted semi-submersibles have been used predominantly. However, there have been a limited number of converted jack-up units placed into service in these areas. Most other applications have been in more moderate environmental offshore areas[4,5] (Fig. 3).

Like a fixed platform, the jack-up unit rests on the seabed during operations, and is not subject to the motions that are experienced on floating semi-submersible, tanker and barge units. Thus far, no new jack-up production units have been built, although designs developed for this purpose have been available for several years.[6]

1.4 Comparison of various types

As demonstrated by past experience, there has been widespread utilization of converted semi-submersibles, tankers, barges and jack-up rigs for marginal field and early production operations. A comparison of the features of these converted types is summarized in Table 1.

However, this paper will concentrate on the design, engineering,

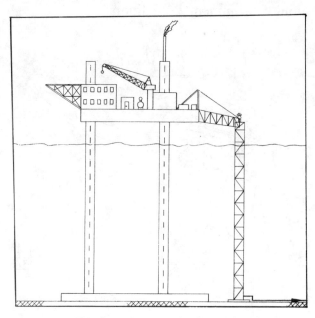

Fig. 3. Jack-up production unit.

TABLE 1

Comparison of Major Features of Converted Semi-submersible Jack-up and Tanker
Production Units

Feature	Semi-submersible	Jack-up	Tanker
Approximate number of conversions (1968–1989)	27	11	12
Available deck space for conversion	2	3	1
Allowable weight capacity	2	3	1
Onboard crude oil storage capacity	0	0	1
Water depth capability	1	2	3
Environmental severity rating	1	2	3
Motion stability on location	2	1	3
Ability to remain on location	2	1	3
Availability of units for conversion	3	2	1
Facility of conversion	3	2	1
Facility of installation	2	3	1
Facility of mobilization	2	3	1
Purchase cost of unit to be converted	3	2	1
Overall cost of conversion	3	2	1

1 = best, 2 = median, 3 = worst, 0 = none.

classification, conversion, installation, operation and decommissioning aspects of converted jack-up rigs for use in early and marginal field production.

In early 1989, a design evaluation was developed for the exploitation of a marginal offshore oil field in the Java Sea.[7] This study will serve as the basis for the discussion presented in this paper.

2 DESIGN CRITERIA

For the selection and conversion design study of an existing jack-up drilling rig for utilization as a marginal field production unit, the following criteria served as the basis:

— field production characteristics;
— water depth at production site;
— environmental criteria at site;
— seafloor conditions at site;
— duration of production service.

2.1 Field production characteristics

The proposed jack-up production unit for the Java Sea offshore oil field under consideration is designed to process the combined effluent from six producing wells. When separated and treated, the effluent constituents are forecast to have daily peak values of 20 000 barrels of oil, 10 mm scf of gas and 8000 barrels of water.

2.2 Water depth at production site

The reference sea-level water depth at the Java Sea offshore site selected for the evaluation study is 65 m. At adjacent areas in the Java Sea, water depths range from 60 to 75 m.

2.3 Environmental criteria for site

Tropical storms or typhoons do not form or track over the Java Sea. Monsoon storms do occur, and account for the most severe weather that is experienced in this area.[8] Northwest monsoons occur in the period from December through February, and southeast monsoons occur in the period from June through October. The NW-monsoon extreme storms are more severe and persistent than the SE-monsoon extreme storms.

The environmental criteria forecast for the most severe 100-year NW-monsoon storm at the Java Sea site have the following maximum values:

— wave height	9·5 m
— storm tide	3·4 m
— crest elevation	9·2 m
— sustained wind	75 knots
— surface current	2·0 knots
— seafloor current	0·5 knots

2.4 Seafloor conditions at site

The seafloor at the Java Sea in the region of the study site is level and flat. Based on the experience gained during exploratory drilling operations, the upper layer of the seabed is determined to be soft but stable mud. Mat-supported jack-up rigs of the type considered in this paper have been used almost exclusively in the region of the site with no foundation bearing problems during normal operations.

2.5 Duration of production service

Production is forecast to be sustained at an economic level for between 5 and 7 years. Drilling and completion of additional wells and workover operations of the initial six wells will be required to maintain that level. The maintenance of the economic production level is forecast on the basis of there not being any provision for the use of gas or water injection for reservoir pressure maintenance or for the use of gas in a gas-lift operation.

3 PRODUCTION SYSTEMS

The production systems selected for installation and operation on the converted jack-up rig are similar to the conventional two-stage separation systems used in the design of other early or marginal field production units.[9, 10] Based on the quantities and properties of the effluents from the producing wells, the production systems equipment is selected to serve the following functions:

— monitoring and controlling of individual wellhead effluent flow and pressure;
— metering, monitoring, control and commingling of the produced effluents from each well;
— testing of the uncommingled effluent from each individual producing well;
— initial separation of the commingled effluent into basic oil, gas and water constituents;
— final separation and treatment to de-gas, dehydrate and stabilize crude oil;
— final dehydration and scrubbing of gas for flare burner disposal and for engine fuel;
— final de-gasing and separation of oil from water for overboard discharge at sea;
— heating, cooling and pressure regulation of separated oil, gas and water;
— metering, monitoring and flow control of separated oil, gas and water;
— pumping and piping between various stages of the production systems.
— pumping of the crude oil into a field flowline system to a storage tanker.

3.1 Production systems equipment

In order to perform the required functions listed above, the production system is designed to incorporate the following equipment:

- wellhead choke manifold
- test manifold
- effluent flow meters
- test separator
- first-stage separators
- second-stage separators
- heaters and coolers
- flotation cell
- HP knockout drum
- gas flare control panel
- gas flare
- fuel gas treater
- crude oil pumps
- produced water pumps
- crude oil flow meters
- production control room
- production laboratory
- fire and gas detection systems
- firefighting systems
- pollution control systems
- personnel safety systems

The above listed production systems features and equipment are selected for marginal field or early production operations in the Java Sea and other similar offshore regions where produced water, treated to international pollution standards (MARPOL), can be discharged overboard, and where gas produced in association with the produced crude oil can be disposed of by flare burning.

3.2 Conversion requirements

In an operating mode, the total weight of the above listed production systems features and equipment is 1000 short tons. This includes the weight of the skid beams on which the equipment is mounted, the interconnecting piping and electrical systems, the weight of the fluids in the system components, and an allowance for other required features.

The weight allowance for the other required features covers the weight of steel for under-deck reinforcement in way of the heavier equipment, additional hull structure, and wellhead support provisions. The weight allowance also includes provision for additional non-production equipment required to operate the converted production unit, and for the future additions of other production equipment.

As a basis for the evaluation study performed for conversion of a jack-up rig to a production unit for Java Sea operations, it was determined that all of the production system equipment would be installed on the open area of the rig's main deck.

The required deck area for the installation of the above listed production systems is 725 m^2, which includes spacial provisions and adequate clearances for the safe operation and maintenance of the various equipment components.

4 CONVERSION RIG SELECTION

The selection of an existing drilling rig for conversion to a jack-up production unit (JUPU) is based on consideration of the conversion requirements for the production systems and the specified design criteria for the Java Sea operating area.

4.1 Rig selection factors

Based on the conversion requirements and design criteria, the following rig selection factors are defined as the following.

— The design capability of the rig is suitable to withstand the forces exerted by the maximum wind, wave and current that can be experienced in the Java Sea.
— The bottom bearing structure of the rig must be suitable for minimum penetration and long-term position and attitude stability while resting on the soft condition of the seabed material.
— The leg length of the rig must be adequate to achieve a suitable air gap (10·7 m) between the storm tide level and the lowest extremity of the upper platform of the rig to allow passage of the crest of the maximum wave.
— The rig must have adequate weight carrying capacity in floating, jacking and elevated modes of operation to accommodate the conversion weight requirement.
— The available open main deck area of the rig must be adequate to accommodate the installation, operation and maintenance of the production equipment.
— The rig must have been built to and maintained in accordance with a recognized classification society.
— The combined effect of the rig's age, its operating history and its existing condition must be such that any modifications required for fatigue strength are minimal, if any are required at all.
— If the considered rig has been inoperative for more than a period of 3 months, suitable provisions must have been made for the preservation, protection and/or maintenance of the rig's equipment, systems, structures, jacking mechanisms and living quarters.
— The rig's AC electric power generator-engine sets must be suitably

rated to handle effectively the demand load of the production systems, the rig utility systems, the living quarters and the jacking system.
— The rig's power generating engines must be suitable for conversion from diesel fuel to produced gas fuel.
— The rig must be available for purchase at a price that is compatible with the economic justification and at a time that suits the schedule requirements for the use of a JUPU in the selected Java Sea Field.

4.2 Selected rigs for conversion

In general, jack-up rigs can be categorized into two types — independent jack-up rigs and mat-supported jack-up rigs. Independent-leg rigs are fitted with individual spud-can structures at the lower end of each leg to develop bottom bearing support for the elevated rig. Mat-supported rigs have a barge-like lower hull to which the legs are affixed at the lower end. The barge-like mat develops the bottom bearing support for this type of rig while standing on location.

Evaluation of the seabed properties and review of the experience gained by oil companies and drilling contractors in the Java Sea area with mobile exploratory and development drilling rigs dictated that mat-supported rigs with their broad bottom bearing area and low compressive bearing load requirement were ideally suited for the intended operation. In other offshore regions, where the seabed is comprized of sand or firm clay, independent-leg jack-up units would be equally suitable.[5]

Jack-up rigs are also categorized by the maximum water depth in which their design criteria and class requirements are satisfied. In the case of the rig to be chosen for conversion, jack-up drilling rigs rated for 75 m water depth were selected for consideration.

The selection of mat-supported jack-up rigs designed and classed for operations in 75 m of water narrowed the consideration down to 28 rigs built by Bethlehem Steel Corp. (Beaumont, Texas) and 6 rigs built by Baker Marine Corp. (Ingleside, Texas). These rigs were therefore selected for further evaluation.

4.3 Evaluation of selected rigs

Jack-up rigs are designed to carry a rated amount of variable load during floating, jacking and elevated operational modes. While jacking, the amount of variable load that can be on board is less than during the other two modes. Since the weight of the installed equipment will be fixed on the JUPU, the jacking mode limit will determine how much permanent weight can be added during the conversion. In the case of the two types of rigs considered, the maximum amounts of weight that could be added for conversion equipment and features plus the weight of the operating supplies and materials are

— Baker jack-up   1000 short tons
— Bethlehem jack-up   1200 short tons

As the above values indicate, the weight limits of the as-built rigs are only marginally adequate to accommodate the weight of the production systems and features to be added during the conversion.

Both types of rig have a drilling slot constructed into the stern structure of the upper platform. Above the drilling slot, the substructure, drill floor and derrick are positioned during drilling operations. During jacking and floating operations, the entire drilling assembly is skidded forward away from the slot.

The main deck open areas available for installation of production systems on the two types of rigs as they were originally outfitted and with the drilling assembly positioned over the drilling slot are

— Baker jack-up   370 m^2
— Bethlehem jack-up   315 m^2

As the above values indicate, there is insufficient main deck open area for the installation of the production systems equipment and features on the rigs as they were outfitted for drilling operations.

Since the producing wells are to be pre-drilled and completed prior to the installation of the JUPU on site, and all major remedial well operations will be performed by a workover rig, there is no vital need for the drilling equipment to remain onboard during the production operations period. Removal of the drilling equipment located on and above the main deck increases the available deck areas on the two types of rigs to the following acceptable levels:

— Baker jack-up   780 m^2
— Bethlehem jack-up   760 m^2

and increases the amount of weight that can be added to the following satisfactory levels:

— Baker jack-up   1500 short tons
— Bethlehem jack-up   1600 short tons

As the available deck area and allowable added weight values show, these two types of mat-supported rigs are comparably suited for conversion for JUPU service. However, since there are a greater number of Bethlehem jack-up rigs from which to select than Baker rigs, the evaluation study at this point focused on the use of Bethlehem rigs for the conversion planning (Fig. 4).

5 TECHNICAL EVALUATIONS

There are a number of design analyses and engineering studies which must be performed in order to determine the overall suitability of the rig selected for conversion. The design analyses to be performed include the following.

— Evaluation of the wind loading on the above-deck portion of the unit with the drilling equipment removed and the production systems added for the floating and on-location modes.
— Site-specific global analyses of primary and secondary strength members.
— Site-specific fatigue analysis covering the forecast service period on location.
— Local analyses of strength members that will support skids and foundation structures of production system equipment and new features.
— Weight distribution analysis of converted rig to determine loads on jacking system during elevating operations.
— Development of hydrostatic curves of form for converted rig if existing drilling slot is enclosed structurally.
— Stability analysis during floating mode with main deck drilling equipment removed and all production systems added.
— Site-specific foundation analysis to verify overturning and attitude stability on site.
— Power load analyses to determine which and how many existing engines on the selected rig will be converted from diesel fuel to gas fuel produced on location.
— Heat radiation analysis of produced gas flare burner flame.

The engineering studies to be performed include the following.

— Study of electrical distribution system to determine required modifications to interface and provide power for production systems equipment and any other new equipment added during conversion.
— Study of piping and pumps systems for well effluent, oil, gas, water, air and hydraulic systems to determine required modifications to interface and provide service for production systems equipment and any new equipment added during the conversion.
— Study to determine hazard zone on main deck when production

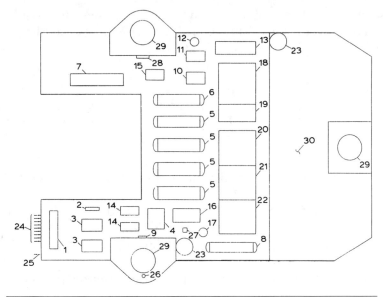

Fig. 4. Production system layout on main deck.

Code	Item	Code	Item
1	Wellhead manifold	16	Utility air compressor
2	Wellhead control panel	17	Utility air receiver
3	Metering skid	18	Switch gear room
4	Surge tank	19	Battery room
5	First stage separator	20	Workshop/storage
6	Test separator	21	Laboratory
7	Flotation tank	22	Control room
8	HP flare KO drum	23	Crane pedestal
9	Flare control panel	24	Flowlines from wellheads
10	Fuel gas treater	25	Crude oil export line
11	Start aircompressor	26	Gas flare (on leg)
12	Start air receiver	27	Utility airdryer
13	Gas engine generator	28	Sub-control panel
14	oil export pump	29	Leg structure
15	Aerial cooler	30	Accomodations

system equipment is operating and the suitability of existing equipment that may be located within the new hazard zone boundary.

— Study to determine the suitability and the required additions or modifications to existing fire and gas detection systems, firefighting systems and lifesaving equipment for production service

— Study to determine the adequacy of the existing pollution containment and waste-treatment systems for production service.

The scope and detail of the above cited analyses and studies are performed in accordance with the rules or guidelines established by a recognized classification society as well as to suit the regulatory governmental rules that may apply for the area of intended service.

6 CONVERSION WORK TASKS

When the completed analyses and studies are reviewed and approved by the selected classification society, the actual conversion of the chosen drilling rig can be undertaken at any suitably located fabrication yard or shipyard.

The overall conversion work tasks can be divided into two groups — the work tasks that focus chiefly on the existing rig structure and equipment, and those that focus on the installation of the production system equipment and features.

The work tasks that involve the existing rig structure and equipment will include the following.

— Rig will be drydocked where outer shells of lower support mat structure can be cleaned and surveyed.
— Any compartments in the mat structure or the upper platform in or near where modifications are to be performed will be gas free.
— Corrosion protection will be achieved by applying protective coatings and by the placement of anodes on and in mat structure and on lower sections of legs to ensure that production unit can remain on location for intended service life.
— All pertinent structure and equipment in mat structure and upper platform will be surveyed to determine compliance with classification society rules.
— All outstanding deficiencies determined by last class survey will be remedied.
— The open area of the main deck will be cleared of drilling equipment and structure, which includes derrick, drill floor, substructure, skidding units, mud treating equipment and bulk storage tanks.
— Main deck and under-deck support structures will be strengthened as required for the installation of production system equipment and features.
— Existing equipment that will be utilized in the converted production unit will be overhauled, repaired, replaced or modified as required by class rules and production service demands.

—Selected engines will be converted from use of diesel to use of produced gas as fuel.

—Revisions will be made to ventilation, exhaust and access openings in main deck in way of area where production systems are to be installed in accordance with classification society approved hazard zone plan.

—Existing lifesaving equipment will be relocated as required in accordance with determined escape routes for the converted production unit.

During the conversion, the work tasks involving the installation of the production systems equipment and features will include the following.

—All skid-mounted and modularized production system equipment will be installed on cleared and strengthened area of main deck.

—Intra-module piping and electric wiring will be fitted and connected.

—Interface connections and fittings will be installed between the rig's existing service systems and the new equipment modules.

—Additional fire and gas detection systems, firefighting equipment, control and monitoring systems and other safety systems relating to the production system will be installed.

—Firewalls and other required heat barriers will be installed in accordance with the results of the heat radiation and fire propagation studies.

—Flare burner will be installed on the top of one of the two aft leg structures.

—Oil spill containment features, waste oil treatment equipment and storage tanks will be installed in accordance with local and international pollution at sea regulations.

—Fabrication and installation of a walkway structure to serve as an access walkway and pipeway between production unit and the wellhead conductor pipe tower.

—Structurally enclose the stern drilling slot to provide added deck space for future installation of production equipment and as storage tank for liquid wastes.

The major events in the project schedule for the conversion of a 250-class Bethlehem jack-up drilling rig to a production unit are listed in Fig. 5 together with a schedule for the conversion of the selected rig in a Gulf of Mexico shipyard and the delivery and installation of the unit at its site in the Java Sea.

```
      MAJOR                           MONTH
    PROJECT EVENT           1..2..3..4..5..6..7..8..9..10..11..


    AWARD CONTRACT          +---------------------------------

    ACQUIRE RIG             ++--------------------------------

    ENGINEERING             ++++++---------------------------

    CLASS REVIEW            --++++++-------------------------

    PROCURE EQUIPMENT       ------+++++++++------------------

    CONVERSION              -----+++++++++++++++++-----------

    MOBILIZATION            ---------------------++++++------

    TOW TO SITE             --------------------------+------

    INSTALL ON SITE         ---------------------------++----

    HOOK-UP SYSTEMS         ----------------------------+++--

    TEST AND COMMISSION     -----------------------------++-

    START PRODUCTION        ------------------------------+
```

Fig. 5. Jack-up production unit conversion schedule.

7 INSTALLATION ON SITE

The field development plan would include the following stages for the installation of the JUPU and other facilities in the field such as the conductor tower, the in-field pipelines, a storage tanker and an SPM mooring system (Fig. 6).

- —A tower structure is fabricated to serve as a support for the well conductor pipes, wellheads, production trees and crude oil export downriser.
- —The tower structure is installed onsite prior to the drilling of the development wells by a mat-supported drilling rig.
- —The producing wells are drilled and completed in advance of the arrival of the JUPU.

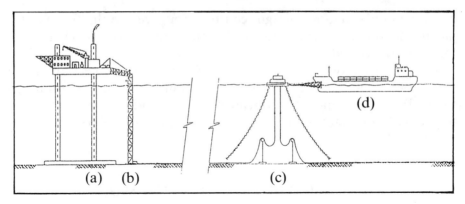

Fig. 6. Java Sea production facilities: (a) jack-up platform 'A'; (b) conductor tower; (c) SPM mooring system; (d) floating storage tanker.

— A storage tanker moored by means of an SPM system or by means of a bow-mounted turret is installed on location in advance of the arrival of the production unit.

— In-field pipelines are installed to link the risers from the storage tanker's SPM underbuoy hoses to the crude oil downriser at the base of the conductor tower.

— When the wells are all completed and the other facilities are in place, the converted JUPU is towed out to and positioned on location according to the operating practices and procedures for a jack-up drilling rig.

— When the JUPU is positioned on location, the walkway structure is extended to span the separation between the conductor tower and the stern of the production unit.

— When the well flow piping installed in the walkway structure is connected to the production stacks mounted on the conductor tower and final testing of all systems is completed, production operations can be initiated.

8 ON SITE OPERATIONS

Once positioned on location, the JUPU is operated in a manner normal for an offshore production facility, except that continued monitoring of the upper platform's attitude and air gap must be performed.

If additional and uniform settling of the mat support structure into the seabed is experienced that decreases the initial air gap, or differential settling causes the upper platform to trim out of level, the leg wedges that carry the upper platform weight can be disengaged so that normal leg jacking operations can be performed to either restore the air gap or level the platform attitude.[11]

The access walkway structure is designed with pin connections at the production unit's stern and at its connection point on the conductor tower. By not having the wellhead platform rigidly connected to the upper platform, slight changes in elevation or attitude can be more easily accommodated.

9 DECOMMISSIONING FROM SITE

When an offshore oil field is depleted by primary and secondary recovery methods and/or production drops below the economic level, a fixed-platform structure remains on the location with no further operational use or economic value. Removing a fixed platform from an offshore site is a costly operation that is by no means offset by the salvage or scrap value of the platform.[12]

The use of a JUPU in a marginal field operation or for an early production operation allows the mobile platform to move off location in the same manner that it did when it was a jack-up drilling rig.

After production at the offshore site is terminated, the storage tanker and its SPM mooring system can be removed and mobilized for subsequent use at some other field location.

The light-framed conductor tower can be removed from the location by use of conventional salvage techniques and a medium-sized offshore construction barge.

The JUPU can either be moved to another field position or mobilized and used in another offshore region where the water depth and environmental conditions are compatible with its design criteria.

In the case where there is no further use for the JUPU, it can be mobilized to a shipyard where it can be converted back to a drilling rig if market demand for jack-up rigs is high, or to a breaker's yard for removal of the production equipment and structural scraping.[12]

REFERENCES

1. Knecht, H. I. & Bernard, S. W., Why mobile rigs can make reliable production units. *Ocean Industry,* **20**(12) (1985).
2. Hammett, D. S., First floating production facility Argyll. Offshore Technology Conference, Paper No. OTC 2821. OTC, Dallas, Texas, 1977.

3. Kobus, L. C. S., Mila field tanker production facility. Baker Marine Technology Report. Baker Marine, Ingleside, Texas, 1985.
4. Cheng, A. L. & Keleher, J. F., First jack-up production platform in North Sea. Offshore Technology Conference, Paper No. OTC 1890. OTC, Dallas, Texas, 1973.
5. Kobus, L. C. S., Converted jack-up rig use as early production unit in Mila field. Baker Marine Corporation Report. Baker Marine, Ingleside, Texas, 1984.
6. Fogal, R. W. & Brown, J. G., Marginal field production system. Baker Marine Corporation Report. Baker Marine, Ingleside, Texas, 1984.
7. Kobus, L. C. S., Converted jack-up drilling rig for use as production unit in Java Sea BGMB (USA) Report. BGMB, Houston, Texas, 1989.
8. Crutcher, H. L. & Quayle, R. G., *Worldwide Climatic Guide to Tropical Storms at Sea.* Naval Weather Service Command, Washington DC, 1974.
9. Kobus, L. C. S., *Multi-use floating production facilities.* Noble Denton and Associates, Inc., Houston, Texas, 1986.
10. Wikholm, L., Kobus, L. C. S. & Salonen, P., Floating production supply vessel for marginal fields. The Way Forward for Floating Production Systems Conference,
11. Kobus, L. C. S. & Whittington, L. V., Jack-up operational guidelines. Offshore Technology Conference, Paper No. 3243. OTC, Dallas, Texas, 1978.
12. Prasthofer, P. H., Platform decommissioning: a look toward the future. *Ocean Industry,* **24**(5) (1989).

The Loss of a Jack-Up Under Tow

A. A. Denton

Noble Denton International Limited, Noble House, 131 Aldersgate Street, London, EC1A 4EB, UK

ABSTRACT

The events leading up to the capsize and total loss of a jack-up under tow are briefly summarised. The investigation of the causes of this are described in some detail. It involved model tests, finite element analysis, stability calculations, a determination of motions by analysis of newsreel videos taken at the time and close examination of an underwater film taken of the wreckage.

The likely sequence of structural failures leading up to the capsize is established. These are summarised and the lessons learnt from the enquiry highlighted.

Key words: jack-up, casualty, capsize, motions, stresses, model tests.

1 INTRODUCTION

The paper describes an actual event which took place a few years ago involving a Marathon LeTourneau type 52 jack-up unit which became a total loss while under tow. Investigations revealed that the design and construction of the platform was adequate for the venture, but that improper towing appeared to be the root cause of its demise.

As a result of the work carried out, a better general understanding of the dynamic behavioural properties and stress distribution patterns in jack-ups has been acquired, and specifically so for the class of unit involved.

2 DESCRIPTION OF THE JACK-UP

Type: Marathon LeTourneau Class 52
Length: 203 ft (61·9 m)
Breadth: 168 ft (51·2 m)
Depth of hull: 22 ft (6·7 m)
Number of legs: 3
Length of legs: 357 ft (108·8 m)
Amount of leg below hull: 10·25 ft (3·1 m)
Displacement: 15 063 kips (6846·8 tonnes)
Draft: 13′–1″ (4·0 m)

3 THE VOYAGE

The 11-year old, fully classed unit, carrying its normal drilling crew, left the shelter of islands in order to commence a 220 nautical mile coastal passage to a lay-up port. Towage was effected by two tug/supply vessels of 70 tonnes and 87 tonnes bollard pull respectively, each acting on a single towline. The rig was towed in a conventional manner from the bow (that is the drilling slot was aft) at 5 to 6 knots. The drilling substructure had been skidded forward clear of the slot. The drilling string and casing was stowed as is normal on the pipe rack between stanchions and secured with chain, wire and turnbuckles. Other deck cargo such as tool bins, portable cabins, etc. were secured by chains or welded stays. It was at or close to its assigned load line.

A forecast was received shortly after the start of the tow which predicted maximum 25 knot sustained winds with decreasing swell for the expected duration of the passage.

Twenty-four hours into the tow the wind was gale force and increasing and the swell increased from 3 m to 5–6 m. Surprisingly the tugs elected not to turn into the weather but to carry on at about 5 knots on a course which put the wind and swell abeam of the tow. Equally surprisingly, perhaps, the jack-up unit was able to ride these conditions with just enough water ingress and minor cargo movement to keep the crew occupied but not to alarm them.

Five hours later the fibre stretcher of one of the towlines broke when the weather was wind 40 knots, sea 3–4 m in 7 s from the west; swell 5–6 metres in 12 s from the southwest. At this time in the words of one of the crew members 'all hell broke loose' as water started to come down the mud return line into the mud pits, and items of deck cargo started to move around significantly.

Immediate steps were taken to secure all watertight doors and hatches that may have been temporarily open, and the majority of the crew evacuated by helicopter. A skeleton crew of nine men was left on board to assist with towline reconnection.

During the next 6 h the unit settled by the stern until only about 1 m freeboard remained aft, then remained in that condition without further settlement. The platform remained basically on an even keel throughout while rolling and pitching about its mean positions at all times. The wind and sea stayed about 40 knots backing slightly, and the swell increased to 6–7 m, still from the southwest. All attempts to reconnect the towline failed, contributed to by the loss of use of the air winches through their flowlines being severed.

The remaining crew elected to abandon the unit as darkness approached and return at first light the following morning in order to attempt to pump out whatever water was present. Two hours later the second towline broke and around the same time the unit capsized and sank in 55 m of water.

4 AVAILABLE EVIDENCE

After the loss the crews of the jack-up and the tugs were interviewed by a governmental enquiry; an underwater video inspection was made of the wreckage; and the broken towing pennants recovered.

The evidence given to the enquiry was inconclusive, and in many cases contradictory as to what had happened after the first towline broke. Later in-depth interviews with the crew of the rig only served to reinforce this position. Thus in order to try to establish what had happened it was necessary to resort to a technical investigation using any other material that could be made available.

The underwater video clearly showed that the jack-up had overturned and was lying keel uppermost on the seabed, with all three legs having broken off. The bow and port legs had broken off above the jackframe. The starboard leg had come away intact along with the starboard aft preload tanks. It could be clearly seen that the bottom plating of the preload tank No. 30 had parted from the box beam at frame 19 (see Fig. 1).

The fibre pennants following recovery were sent away for testing (this is discussed further in Section 6). It was very clear however that they did not have hard eyes and that failure in both cases had been in the eyes. Other evidence that became available was extensive footage taken by local television companies from fixed wing aircraft and helicopters. This footage was from the time that the majority of the crew were being

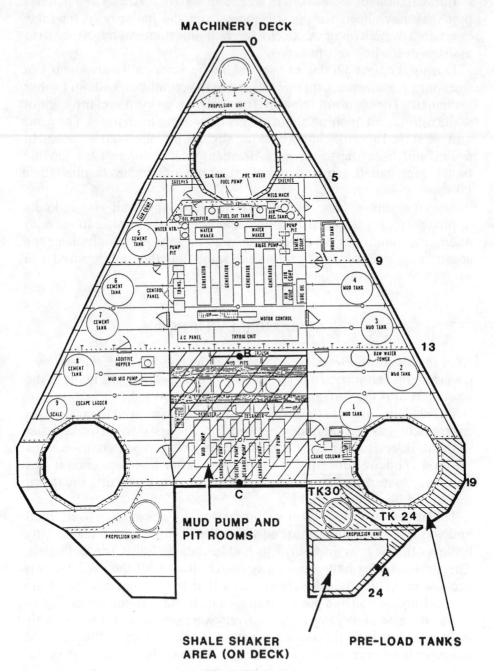

Fig. 1. Layout.

evacuated following the first towline break, and then later at about 2 h before the final evacuation.

From this footage the motions of the stricken unit could be observed, and it could also be seen that it had a pronounced trim by the stern, particularly at the time of the second footage. It was also possible to determine the headings of the unit in relation to the wind, sea and swell.

Also in the area were airforce rescue helicopters who took many high quality still photographs. From these it was possible to establish that no heavy components of deck cargo had moved, which pointed toward the stern trim being due to water ingress rather than cargo shift. This confirmed the views of the crew members who remained on board, but they were unable to identify with any precision where, when and to what extent water had entered the jack-up. It was also possible from the photographs to accurately measure the size of towing chains and wire pennants deployed, as these dimensions were also the subject of substantial variation in the crew's evidence.

5 THE ANALYSIS

5.1 Consideration of leg length

From the evidence it could be seen that the unit in its intact condition could survive in the weather conditions that prevailed. As further corroboration of this point calculations carried out showed that to overturn this class of jack-up floating at its assigned load line there would have to be acting simultaneously a 70 knot wind and a 21 m significant sea, which clearly did not apply in this case. So how did water enter? Firstly in order to give the unit a trim by the stern; secondly such that the trimming ceased after a few hours; and thirdly in order to cause capsize.

Ways in which water could enter were by cracks in the deck, cracks in the side and bottom shell plating, through pipes penetrating the hull, and through hatches and ventilators.

The crew had observed cracks in the deck close to the jacking gear boxes, and suggestions were made that the legs were too long for the tow. To check this point model tests on a scale of 1 : 60 were carried out in a commercial model basin using a fully instrumented model. Measurements were made of shear forces and bending moments in the longitudinal and transverse directions on one of the legs. These were then vectored and the vectored channel searched for instantaneous maxima.

A finite element stress analysis was also made using PAFEC, in which the limiting allowable leg capacity was determined. The result is shown in Fig. 2. The leg analysis was carried out using a structural model of the leg and jackframe assuming the bottom of the jackframe to be rigidly supported. Pinion stiffness and pinion backlash values based on actual measurements made on board the unit a few months before the loss were used.

The method used to combine axial and bending stresses in the leg members is that given by the American Institute of Steel Construction *'Manual of Steel Construction'* 8th Edition. Allowable stresses used were 0·88 × yield stress in bending, 0·8 × yield stress in tension. Purely rotational motions about the centre of flotation were considered, i.e. heave was ignored, and no increase to account for shock loading was made as the legs were secured by the proprietary clamp bars designed

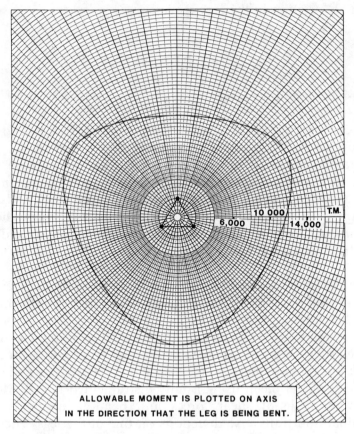

Fig. 2. Allowable leg moment (allowable BM at upper guide based on full leg length, 10 s period).

and fitted by the manufacturer. The limiting allowable leg moment is obtained when the leg is loaded towards a chord. It is produced by a roll or pitch of 11·3° in 10 s.

Combining the results of the model tests and the leg analysis it was found that a general limiting sea state for this class of unit with full leg length for application to tows is about 6·5 m. If one takes off one leg section of 33 ft there is a reduction in motion induced stresses in the legs of about 35%.

The model tests addressed specific sea states and headings for the unit that occurred throughout the tow up to the time of the loss. No moments were induced which exceeded allowables, so the legs did not appear to be a cause of the problems. In the model tests the ongoing weather conditions that would have been experienced had the tow not been interrupted by the first towline parting were also applied. Again no moments in excess of allowable would have been seen.

5.2 Motion induced stresses in the hull

A finite element model was then constructed of the starboard leg jackframe, leg well and adjacent hull which included the starboard aft preload tanks (see Fig. 3). Note that the model basin tests also included

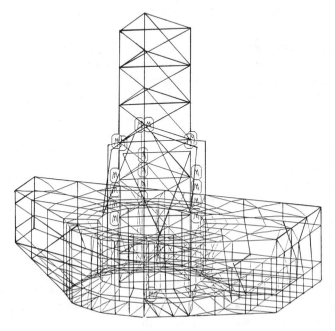

Fig. 3. Complete structural model.

provision for measurements of shear force and bending moment in the hull at frame 19. Firstly, the allowable leg bending moments together with relevant shear forces and vertical loads were applied to the numerical model in six directions. Secondly, leg and hull forces and moments derived in the model tests were applied.

The results of the analyses were that no significant hull plate overstressing (i.e. above 80% of yield) or possible modes of hull failure could be found. It confirmed that the hull can be regarded as being stronger than the legs, such that the legs ought to fail before the hull. This is as prototype experience shows. Also, that on the tow in question the hull was not overstressed as a result of loadings from the legs and/or sea induced flexure.

There were however indications of stress concentrations in the main deck in the region of the jacking gear box to deck connections such as could cause cracks, but not sufficient to provide a mechanism for hull failure. As already noted such cracks were in fact reported by some crew members, being hairline in width and 12–18 in (30–45 cm) long. They were not sufficient to permit significant amounts of water to enter the hull. Thus the entry of water from hull cracks induced by leg moments and hull bending could not be the cause of the loss, so attention was paid to hull penetrations.

5.3 Water ingress through hull penetrations

Crew evidence indicated that before evacuation there had been no significant water ingress through hull penetrations, except via the 14 in. diameter mud return line. The mud return line connected the sand trap, itself located below the shale shakers, to the mud pits in the hull. There was no means provided of closing this line.

For water to enter the mud return line it would be necessary for waves to strike the shale shaker and thus run into the sand trap. The sand trap and shale shaker were located as shown in Fig. 1, surrounded by a wall 2·8 m above the main deck and 5·4 m above the still water line, these being the heights water would have to reach to fill the sand trap and thus flow down the mud return line.

Before the towline broke, when the unit was being towed at 4·7 knots beam on to the weather, only a small amount of water had come down this pipe, and was easily handled by a mud pump. Immediately after the towline broke the flow rate through the pipe increased dramatically such as to lead to overflowing of the mud pits. At this point the crew abandoned the mud pump and pit rooms, closing the watertight doors of the pump room behind them, and thereafter it was impossible for them to tell how much water entered these rooms.

In the model tests the height that the water reached at three points on the deck was measured. These locations were at the shale shaker, abaft the accommodation, and over the mud pump room hatch just forward of the drilling slot. They are shown as A, B and C on Fig. 1. In the situation in which the unit was being towed at speed, there was only occasional overtopping of the shale shaker. At locations B and C there were 1·6 m and 0·5 m of water over the deck.

When towing ceased, as appeared to happen following the first towline failure, the heights of water over the deck were:

Shale shaker	6·7 m
Abaft accommodation	3·6 m
Pump room hatch	3·4 m

These amounts of water were easily sufficient to start dislodging cargo such as crates and skips and, very significantly, to dump large quantities of water over the shale shaker. The crewman's graphic description of events given earlier would not be surprising in the circumstances.

The model tests clearly illustrated the beneficial effect that tow speed had on effective freeboard. The change from 4·7 knots to zero caused water levels over the deck to increase between times 2·3 and 6·8, depending on location. A test was made to establish water levels for the case of the tow being head to weather, which would have been a normal attitude to adopt in gale conditions. Here the water heights were:

Shale shaker	1·6 m
Abaft accommodation	Nil
Pump room hatch	Negligible

Clearly in this situation the deck cargo was protected and there was no threat of water entering the mud return line.

That water would not overtop the shale shaker when the unit was head to weather was further confirmed by the newsreel footage taken about 4 h after the towline break. Here the unit was clearly seen heading into the prevailing weather with no overtopping of the shale shaker. A small trim by the stern was also to be seen. Later footage taken 7–8 h after the towline break also showed the unit head to weather, but now with major settlement, having only 1 m freeboard left aft. It was about this time that it was seen by those watching the stricken unit that its settlement in the water had halted.

The amount of water which was calculated to have flowed down the mud return line before the unit came head to weather was insufficient to cause the substantial stern trim seen later. The trim had also increased during a time when no water was entering the sand trap. Thus it was concluded that there had to be water entering the hull from a location other than the mud return line.

It was suggested that water could have been entering the aft preload tanks. It is certainly not inconceivable that a split could have occurred on one of the welded joints of the shell plating as this has been known to have occurred before on older units subjected to heavy weather. However splits of equal size would have had to open simultaneously in both port and starboard tanks, since the unit remained virtually on even keel throughout. Also they would have had to close up again at some point if the cessation of the settlement into the water was to be explained. This theory was discarded as being impractical.

5.4 Determination of static heel and trim

In order to be able to calculate the amount of water in the vessel at the time the settlement ceased, as was desirable if the analysis was to progress, it was necessary to know the static heel and trim. Fortunately the newsreel video taken from a helicopter contained sufficient continuous coverage to permit analysis of the motions of the unit from which the static condition could be determined.

The signal from the video recorder was digitised to produce a picture of 512×512 pixels. This permitted a film frame to be frozen with complete steadiness and clarity, and to be enlarged and enhanced as required. From this still the instantaneous position of the platform relative to the horizon and the helicopter could be ascertained. By taking a series of frames in sequence over some cycles the motion of the rig was analysed. To permit this analysis complex equations were developed which had to take into account the two-dimensional view of the image from a moving object and the full three-dimensional description of the platform all in an arbitrary orientation. Conditions applying at the time were:

Weather: Wind WSW 40 knots
            Sea WSW 3–4 m, 7 s
            Swell SW 6–7 m, 12 s
Heading: 248°

The inclination of the unit induced by the wind was calculated from a knowledge of the geometry of the unit and checked by the model tests. These inclinations were zero heel and 0·8° by the stern.

The solution to the equations mentioned showed the unit to be rolling

2° each side of vertical in 12·5 s, and pitching 5° also in 12·5 s. The mean angle of heel was 0·3° to port and the mean angle of trim was 3·5° by the stern. After adjustment for the contribution from the wind moment the static conditions were found to be 0·3° to port and 2·7° by the stern.

Before abandoning the vessel the crew had entered the rooms on each side of the abandoned pump room, and from their evidence it was estimated that there was a net amount of water on the port side of the amount required to provide the static heel seen in the newsreel video. It was also thought that a relatively small amount of water had entered the aft preload tanks through the access lids to the manholes when the unit was exposed to following seas. These amounts were thought to be the same in both port and starboard preload compartments, but insufficient to cause much trim. Thus the evidence pointed towards there having to be a large amount of water in a centreline compartment to cause the observed trim by the stern. Clearly the only compartment which could provide this was the combined mud pump and pit rooms.

A calculation was run on a GENSTAB[1] computer model which showed that the amount of water required in the referenced rooms to cause the observed trim was that which they would have if the hatch over the pump room was open and the rooms had filled to this level. Support for this view came also from the fact that once full to this level no more water would enter and settlement by the stern would cease — as indeed happened. Also this particular hatch was virtually at deck level. It had only a 100 mm (4 in.) coaming to permit it to be below the pipe racks. Thus water washing over the decks, as it did for much of the time, could continuously enter the compartment even after it had stopped entering via the mud return line.

5.5 Overloading of the pump room hatch

So the question had to be asked why had the hatch opened. Crew evidence was again not too helpful except to point towards the hatch leaking nuisance water before the towline broke and possibly being distorted afterwards. However, photographs taken at the time of crew evacuation showed high seas breaking over the deck, and confirmed in the model test results given above. These findings led to a finite element stress analysis on the hatch cover being carried out to check for its ability to take high hydrostatic pressures.

The hatch cover configuration is shown in Fig. 4. Its main elements are $84'' \times 84'' \times 5/16''$ plate stiffened by one $4'' \times 4\frac{1}{2}'' \times \frac{3}{8}''$ and two $2'' \times 2'' \times \frac{1}{4}''$ T-bars in the longitudinal direction and three $2'' \times 2'' \times \frac{1}{4}''$ T-bars in the transverse direction. A $1\frac{1}{2}'' \times \frac{5}{16}''$ flange ran all round the edge of the hatch cover.

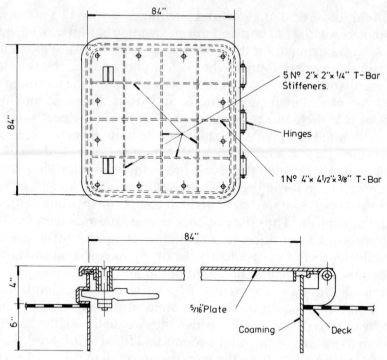

5 № 2"× 2"× ¼" T-Bar Stiffeners.

Hinges

1 № 4"× 4½"× ⅜" T-Bar

⁵⁄₁₆"Plate

Coaming Deck

Fig. 4. Hatch stiffening (1 in = 2·54 cm).

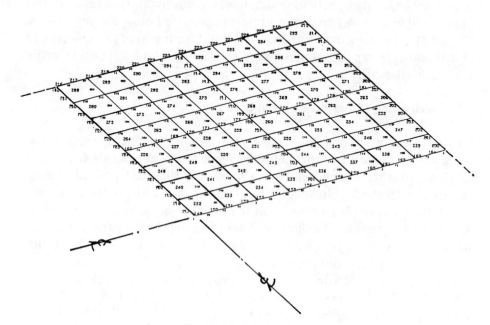

Fig. 5. Model of hatch.

The PAFEC model is shown in Fig. 5 and consists of eight-noded quadrilateral elements capable of carrying pressure loads normal to their surface together with offset beam elements to model the T-bar stiffeners. Symmetry conditions were used which allowed the model to consist of only one quarter of the hatch. The edges of the hatch were modelled as being rigidly supported in the vertical direction, but with no edge moment or in-plane restraints.

The study showed that at a pressure of 5·3 tonnes per square metre (i.e. 5.2 metres of water head) the flange of the main T-bar begins to yield, resulting in permanent distortion providing the dogs are not in place. At this applied pressure it was calculated that the gasket would compress sufficiently to release the pretension in the dogs and permit them to walk back. The pattern of permanent distortion of the hatch cover is shown in Fig. 6. There is a tendency for the rim of the hatch cover to lift away from the coaming to the same order of magnitude as the applied plastic distortion at the centre of the panel.

The analysis carried out was valid irrespective of the condition of the plating and stiffener welds as the hatch cover strength relies principally on the main T-bar stiffener, and stresses in the welds are very low. The model tests showed that when the first towline broke and the tow stopped, the water head over the pump room hatch was 3·4 m, insufficient to damage it. It will be recalled that at this time the jack-up was beam on to the weather.

The model tests also showed that when disconnected from the tugs the jack-up would be broached to the sea, occasionally kicking its bow up to face a particularly high wave then settling back to a broached position. It would never try to go stern to weather. This pattern of behaviour is consistent with the author's own prototype experience.

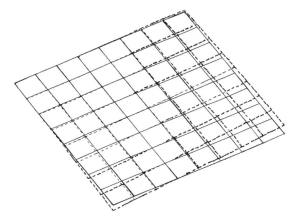

Fig. 6. Quarter hatch with rim residual distortion. No dogs restricting vertical movement.

However, suppose the tug still attached had rotated the unit to put it stern to weather. The model tests showed that with stern up to the prevailing weather, heads of water of 6·2 m could be attained over the pump room hatch, sufficient to distort it. It became necessary therefore to establish whether the unit ever became stern to weather.

5.6 Orientation of the jack-up unit

As before evidence from those present was confused about the orientation of the unit shortly after the towline broke, except to say they were either ahead or astern to the weather! Thus evidence had to be found elsewhere. This came in the form of observations made in the control room of the pitch of the unit and logged in a notebook that was brought ashore.

The control room observations were that the unit, which at that time had not started flooding to any extent such as to trim it from level, was pitching forwards 10° and backwards 15°, and rolling only slightly. This clearly indicated that it was indeed at or close to ahead or astern to weather.

Model tests were carried out for the prevailing weather in which pitch was measured for two cases, head to weather and stern to weather. When head to weather the model pitched 13·2° forward and 9·9° aft. When stern to weather it pitched 11·2° forward and 14·2° aft, which matched the prototype measurements. Thus it was not only shown that the unit had to have been stern to weather, but also that this type of unit tends to pitch towards the oncoming seas. The latter feature again confirms the author's prototype experience. It is also of interest to note that during the towing tests at 4–5 knots the model had a tendency to incline towards the weather, whether it be beam on or from starboard aft quarter.

The picture of what happened up to the crew finally abandoning the platform now seemed to be in place. Initial flooding occurred through the mud return line when way was lost as the towline broke, and was some time later stopped by turning the unit head to weather. Then further flooding occurred by reason of turning the jack-up stern to weather very soon after the towline broke resulting in seas coming on board such as to distort the hatch cover over the mud pump room. However once the pump and pit rooms had filled no further water could enter, and the unit still remained afloat. So why did it sink a few hours later?

5.7 Overstressing in the hull

Reference was made earlier to the underwater video and the fact that the starboard leg had come away intact together with the preload tank. This clearly indicated a weakness in the hull at frame 19, otherwise the starboard leg would have snapped off above the jackframe as it hit the seabed during the overturning process, as in the case of the other two legs.

Attention was now paid to the construction of the unit at frame 19 by means of a further finite element stress analysis. Global analysis as mentioned above did not give rise to any concerns, the hull not being overstressed from leg moments or from its own longitudinal flexing. However the model tests showed that when stern to weather shortly after the towline broke very large heads of water must have been experienced. At the shale shaker and pump room hatch they were about 6·3 m over the deck which corresponded to 13 m at the bottom plating, enough to give rise to concerns about the possibility of local overstress on the bottom shell construction.

A detailed PAFEC model was constructed of the bottom plating, longitudinal stiffeners and end connections in the region of frame 19. The details of the construction and the model are shown in Figs 7 and 8 respectively.

Vertical pressure was applied to the bottom plate and it was found that the greatest stresses were in the connection between the 9″ × 9″ bracket and the stiffener. Yield would occur in unwasted material at a sea water pressure head of 10·6 m, and at a smaller head if corrosion were present. It was deduced that these connections failed as a result of the high pressure heads induced by having the stern up to the weather. (Had the unit been kept up into the weather the head on the bottom plating would have been about 7 m, insufficient to cause problems.)

The failure of the stiffener to bracket connections still left the bottom shell plating attached to the box beam so water ingress was not immediately possible. However a re-run of the PAFEC model for this new unsupported condition showed that stresses in excess of twice yield were occurring in the bottom plating, and this of course from cyclic loading. A low cycle fatigue failure mode was therefore postulated for the bottom shell plating with cracks growing from the toe weld between stringer and bottom plating until the plating split open. The time frame of about 12 h from first overload of the bracket to stiffener connections to failure of the bottom shell was compatible with other prototype experience.

Once the bottom shell split, rapid ingress of water into the starboard preload tank occurred, resulting in overturning in the direction of the starboard leg. At this time the strength of the hull near frame 19 starboard was considerably reduced which permitted the preload tank to tear away and bring the leg with it. It is of course a thankful miracle that the crew had decided to abandon the unit just two hours before it capsized.

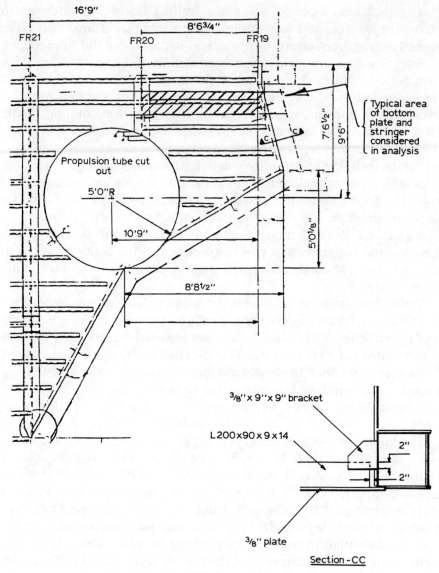

Fig. 7. Stiffening for bottom plate (1 ft = 0·3048 m, 1 in = 2·54 cm).

6 INCIDENTAL FINDINGS

In addition to providing some of the answers relevant to the specific loss investigation, the model tests also gave other useful information about the behaviour of this particular design of jack-up. If one round of leg (33 ft) is removed, as noted above, the sea induced motions in the legs are

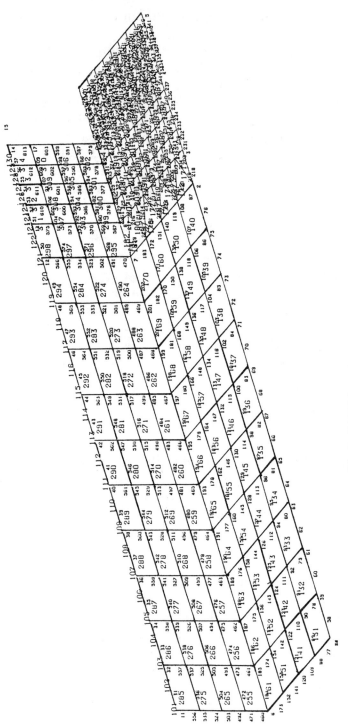

Fig. 8. PAFEC model of stiffened panel.

substantially reduced. However the amount of water over the deck aft is not significantly altered and in some instances is worse. The motions of this type of unit are not much affected by quite substantial changes in the radii of gyration. The degree of leg bending moment induced is very much more related to the height and weight of leg.

The model tests also showed how leg moments can be reduced when head to weather by letting the unit fall back before the weather. In a 9 m sea a moment reduction of 30–40% can be obtained in this way.

Tow resistance was measured in the model tests. Excellent correlation was obtained between the force to hold station head to weather using the standard in-house method of calculation and that found in the tests for a variety of sea states. It was found that allowing the model to drift back while keeping head to weather could reduce the towline forces by over 50%.

The fibre stretchers used in the towlines were laboratory tested which yielded useful information. They were constructed from $11\frac{1}{2}''$ circumference 8-strand cross plait 6,6 nylon rope and made up into grommets with soft eyes. This type of grommet properly terminated would have a new dry strength of about 215 tonnes, and a wet strength of about 190 tonnes. The tests showed that it would lose about 4·4% of its strength per day in the kind of towing service described in this paper. The effect of using a soft eye over a shackle pin was to further reduce the strength of the grommet by 33%.

7 CONCLUSIONS

The investigative analysis described in this paper was in itself of local and specific interest only, although of considerable stimulation to those concerned at the time. What is of general interest is that its pursuit revealed a great deal about the Marathon LeTourneau 52 class of jack-up. It was shown that this design rides comfortably and safely at sea providing it is treated properly. When head to, or partly head to, heavy weather the motions can be acceptable for carrying full leg length, the deck cargo is protected and no excessive stresses are seen on the hull.

Even if the towlines break the design will remain at worst beam on to the weather when motions will still be acceptable and the hull stays satisfactorily stressed. However it is important that in this situation cargo is well tied down and the mud return line blanked off. What the design is not meant for is deliberately putting it stern to heavy weather, which incidentally in the author's experience is also true for other jack-up designs.

In extremes the legs will fail before the hull, a feature noticed from prototype experience, and in the circumstances to be preferred. With full leg length, sea states up to 6·5 m can be contemplated without overstress. The leg critical motion curve actually passes through 11·3° and 10 s providing leg clamps are fitted, but for practical purposes it would be better to continue with the long established 10° and 10 s limit and retain some reserve. The sea induced motions of the design are decidedly non-linear and it does not lend itself to current computer motion analysis techniques.

When towed at over 4 knots, beam on to the weather, there is a noticeable tendency for motions to be dampened and less water to come on deck. However this is not recommended in heavy weather as the penalty for towline failure is to find oneself badly broached which will induce substantial water over the deck. As indicated earlier it is best to keep this type of unit heading into heavy weather, or say not more than four points off as dictated by optimum comfort in the weather prevailing. In these positions it will be a well found vessel.

ACKNOWLEDGEMENTS

The investigations into the loss described in this paper were undertaken by a team comprising D. A. Ipp QC, K. J. Martin and C. D. Steytler of Parker & Parker, and the author with close support from P. M. Lovie of Lovie & Co. Invaluable contributions were provided by R. V. Ahilan, R. B. Bush, N. Lynagh, R. B. Matten, R. W. P. Stonor and N. L. Wilton of Noble Denton; J. Campbell of Ceanet, I. J. Collins of Arctec, L. R. Crosse of MOCL, I. W. Dand of BMT, M. R. Parsey and A. Ratcliffe of Tension Technology, and R. J. Perkins of Cambridge University.

REFERENCES

1. *Users Manual: GENSTAB*, Noble Denton Intl Ltd, London, 1985.

Fatigue of Jack-Ups: Simplifying Calculations

E. C. Hambly & B. A. Nicholson

Edmund Hambly Ltd, Home Farm House, Little Gaddesden, Berkhamsted,
Hertfordshire, HP4 1PN, UK

ABSTRACT

Simplification of calculations for fatigue of jack-ups is essential in order to understand the primary influences and the uncertainties involved. Considerable simplification is possible for jack-ups because calculations for fatigue are dominated by the few worst loading conditions, while the large number of lesser conditions can be ignored. The dominating conditions are likely to be — during operations, the resonant sway and surge motions at the natural period of the structure; and during towage, the 1 or 2 days of maximum roll and pitch motions. Examples for both situations examine loading conditions and demonstrate simplified calculations.

Key words: fatigue, jack-up, platform, offshore, operations, towage.

1 INTRODUCTION

A rigorous fatigue analysis for a mobile jack-up platform is a horrendous matter to contemplate. Since real conditions cannot be predicted precisely, even for known locations and configurations, there is little point in trying to calculate fatigue damage in every detail for every possible life history. It is essential to make major simplifications in order to understand the primary influences and to identify sources of error. This paper illustrates some of the complications and indicates how fatigue calculations can be focussed on critical conditions.

Example calculations are presented for operating conditions of the two idealised jack-up platforms illustrated in Fig. 1. It has been assumed

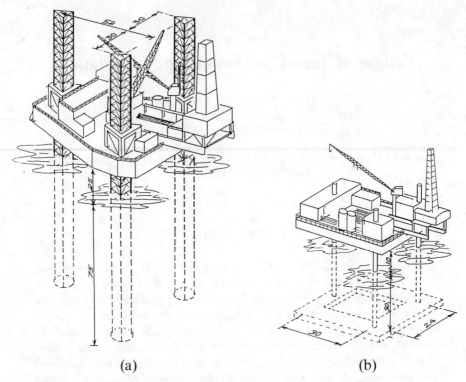

(a)   (b)

Fig. 1. Jack-up examples: (a) independent leg unit in 75 m water; (b) mat-supported in 30 m water.

for the independent leg unit that there is significant spudcan moment fixity under operating conditions, as has been observed on location.[1] As a result there are moments at the leg/spudcan connection, and a detail in this region has been adopted for the sample fatigue calculations.

The bottoms of the legs will be subject to fatigue loading on all locations. Critical regions of legs further up must also be checked, particularly if they are likely to be located between the upper and lower guides of the structure for a significant part of their life. The jack housing and parts of the hull will also be subject to fatigue loading at all locations.

During transit the bottoms of the legs and supporting regions of hull are subject to fatigue loading by the associated roll and pitch motions. As above, major simplification of calculation is warranted because the real conditions cannot be predicted with accuracy. Also, since fatigue damage is so extraordinarily sensitive to the height of the legs during the tow, relatively small changes to leg height can change the calculated fatigue damage from unacceptable to acceptable by a wide margin.

A considerable amount of advice on fatigue and jack-ups is included in Refs 1–12 (and their references). This paper is not intended to be a review or guide to the subject, but presents a number of specific observations and calculations.

2 OPERATING CONDITION

2.1 Natural period of unit

A jack-up elevated in the operating condition is subjected to continuous oscillations at its natural periods by the random waves and winds. For most of the time these oscillations are not noticed by men on board: however, even in fine weather they may be identified by small movements of the levelling bubbles. During stormy weather the oscillations at the natural periods are superimposed on the motions of the slower period associated with the storm waves, as illustrated in Fig. 2.

The sensitivity of a jack-up to resonance at its natural periods is illustrated in Fig. 3. Figure 3(a) shows the range in sideways bending moment ΔM_x induced at the bottom of the forward leg of the independent leg unit by waves of various periods and heights. Figure 3(b) shows the equivalent moment range for the mat-supported unit. In both diagrams the size of the wave has been increased with period towards the right in order to be representative of the steeper waves observed at each period. It is evident in (a) that because of resonance a small wave of 1·6 m × 4·7 s can cause a moment range of the same magnitude as a large wave of 10·5 m × 10 s. Since very many more small waves occur than large waves, the oscillations due to resonance in the small waves are likely to cause much more fatigue damage.

The natural periods of a jack-up can be calculated by a dynamic stiffness analysis computer program, which is virtually essential at some stage of a fatigue analysis for a jack-up. The natural periods in surge and sway are given approximately by

$$T_s = 2\pi (M/\Sigma K)^{0.5} \tag{1}$$

where

T_s = natural period in surge or sway,

Fig. 2. Deck motions with resonance at natural period.

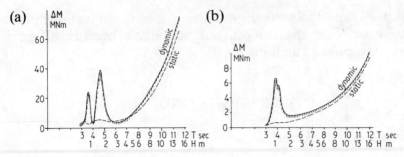

Fig. 3. Moment range at bottom of forward leg due to waves of increasing size from the beam: (a) independent leg unit example; (b) mat-supported unit example.

M = mass of hull and upper regions of legs which surge/sway with the hull,

ΣK = sum of sway stiffnesses of legs in direction of motion (including P-delta effect).

The independent leg unit example is found to have a sway natural period of $T_s = 4\cdot7$ s, while the mat-supported unit example has $T_s = 3\cdot8$ s.

Jack-ups can have significant yaw oscillations if the natural period in torsion is large. The yaw natural period is also calculated by the dynamic stiffness analysis: it is given approximately by

$$T_y = 2\pi(Mr_m^2/\Sigma(Kr_k^2))^{0\cdot5} \qquad (2)$$

where

r_m = radius of gyration of hull mass M,

r_k = radius of leg of stiffness K from centroid of legs.

In the independent leg unit example $T_y = 3\cdot6$ s, and it is evident in Fig. 3(a) that significant moments can be induced by waves at this period. The mat-supported unit has $T_y = 4\cdot2$ s. T_y is less than T_s for the independent unit because most of the hull is in an area between the legs; i.e. r_m is less than r_k. In contrast, the mat-supported unit example has the quarters forward of the forward leg, so that r_m is greater than r_k and the yaw oscillations are slower than surge oscillations. In a random sea-state, surge, sway and yaw oscillation are induced simultaneously by wave components of appropriate periods. If the natural periods are close to each other the resonances beat with each other while their phasing changes[2] and a detailed analysis may be required of their interaction.

Figure 3 has been drawn for wave loading from the beam because the structures are not symmetric with respect to such loading and yaw motions are induced with sway.

2.2 Wavelength and direction

The response of a jack-up to waves of short wavelength is very sensitive to the direction of wave approach. Figure 4 illustrates with 'roses' how the moment range in sideways bending of the forward leg of the independent leg unit example depends on the direction of waves. Rose (a) shows the moment ranges due to waves of 0·8 m × 3·6 s × 20 m from different directions, (b) and (c) show the moment ranges due to 1·6 m × 4·7 s × 34 m waves and 11 m × 10·5 s × 170 m waves, respectively. The moment ranges fall to zero between the rose petals when the yaw and sway motions at the forward leg cancel each other. It is evident in (a) and (b) that a change of angle of approach of 5° can make a difference of tenfold in the moment induced. Figure 4(d) shows how the crests at 34 m spacing coming from 92° cause simultaneous loading on all three legs, while (e) shows how waves from 110° load the legs in opposite directions so that their effects largely cancel out. It is shown later that most of the fatigue damage for this unit is caused by waves at the resonant periods. A rigorous fatigue analysis would need to study the wave directions in great detail, possibly involving 100 or more lines of approach for some sizes of waves. Fortunately, fatigue damage is proportional to the stress to power of 4, and consequently 90% of the lines of approach in (b) cause negligible fatigue damage and can be ignored as compared to the 10% of wave directions in the dominant 'petals'.

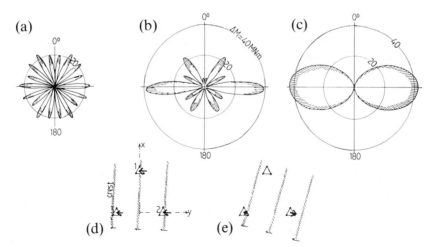

Fig. 4. Sensitivity to wave direction of moment range of ΔM_x at bottom of forward leg of independent leg unit example: waves of (a) 0·8 m × 3·65 × 20 m, (b) 1·6 m × 4·75 × 34 m, and (c) 11 m × 10·55 × 170 m; (d) and (e) synchronous and non-synchronous loading of legs by waves from slightly different directions.

The number of petals in the response rose depends on the interference between the wave crests and the legs of the unit. Waves of shorter length produce more petals in the response rose, while waves of long wavelength only produce two petals. Since the wavelength of 11 m × 10·5 s waves is much greater than the leg spacing, all three legs are loaded by the same wave at the same time, and rose (c) has only two petals with large moment ranges for waves from about 25% of directions.

Figure 4 indicates that maximum sideways bending moments in this example are caused by waves coming from the beam at 90° or 270°. It is not uncommon for the direction of response to be quite different from the directions of the waves. In the mat-supported unit example at resonance in small waves, it was found that maximum lateral bending of the forward leg was caused by waves from 60°, 110°, 250° and 300°. The overall spread of critical directions then amounted to 15% as compared to 10% in Fig. 4(b).

The roses in Fig. 4 were produced with a dynamic modal analysis program with three degrees of freedom (surge, sway and yaw) and legs represented as single members of equivalent flexural and hydrodynamic properties. Linear wave theory was used because it was found to differ little from higher-order wave theories for the small waves which cause fatigue damage. Waves were assumed to be long crested.

2.3 Dynamic amplification and damping

The dynamic response of a jack-up near resonance is very sensitive to the dynamic amplification factor, which in turn is very sensitive to the damping. The dynamic amplification factor (DAF) of a single-degree-of-freedom system at resonance in a steady stage is given by

$$DAF = 1/2c \qquad (3)$$

where c = damping ratio, as fraction of critical damping.

Monitoring of jack-ups on location[1] has shown that c can be as low as 2% in mild sea-states when analysed by spectral methods. Similar levels of damping have been observed on fixed platforms.[3] In the deterministic calculations reported in previous paragraphs, damping of c = 5% has been used. This value has been adopted because oscillations do not as a general rule build up as a steady state. In a transient analysis of a structure at resonance with 2% damping, the DAF increases to about 6 after two cycles, 10 after four cycles, 16 after eight cycles and 25 in the steady state. A DAF of 10 is considered to be representative of severe resonance, and corresponds to a steady-state analysis with 5% damping.

However, under average conditions the resonant oscillations continuously grow and decrease again and it is assumed for the deterministic fatigue calculations in Section 2·5 that an average DAF of about 7 is relevant, corresponding to steady-state damping of 7%. In spectral analyses, referred to in Section 2·7, 2% damping is more appropriate.

2.4 Wave data

A fatigue analysis for a jack-up needs information on the numbers of waves of each height and period, particularly for waves with periods near the natural period of the structure. Part (a) of Table 1 shows the numbers of waves in each 2 m range of wave height between 0 and 12 m, and each 1 s range of period between 3 and 12 s. This table was derived from the distributions of wave height and period prepared for fatigue analyses of fixed structures. Unfortunately the data are not precise enough for an accurate analysis of a jack-up. It is shown in the next section that a large part of the calculated fatigue damage is associated with the highest waves at the natural periods of the unit. In the case of the independent leg unit example these waves are in the boxes for periods 3–4 and 4–5 s. The number of waves estimated for these boxes is very sensitive to the mathematics of extrapolation of distributions of waves between heights and periods, because they represent wave sizes at the edges of the distribution. Reasonable confidence might be placed on statistics prepared for waves in the middle of the distribution, such as for 5 m × 8 s; much less confidence should be placed on statistics at the edges.

2.5 Fatigue damage

In the following paragraphs the distribution of fatigue damage with wave size is calculated for a detail at the bottom of the forward leg of the independent leg unit example. Damage is first calculated for waves with periods much greater than the natural period of the unit, and then calculated for the particular conditions at resonance.

Part (b) of Table 1 indicates the nominal stress range calculated for the mean wave in each box of wave size (except for conditions near resonance discussed below). For example, waves of 11 m × 10·5 s induce a stress range of 49 MPa. If it is assumed that the stress concentration factor (SCF) is 3 at the detail under consideration, the hotspot stress range is $\Delta\sigma = 3 \times 49 = 147$ MPa. The American Petroleum Institute (API)[3] S–N curve X indicates that the fatigue damage per wave cycle is

TABLE 1

Wave Data and Calculated Fatigue Damage for Jack-Up on Location Offshore Laputa

(a) Numbers of waves of each height and period

Wave height (m)	Wave period (s)								
	3–4	4–5	5–6	6–7	7–8	8–9	9–10	10–11	11–12
10–12	—	—	—	—	—	1	10	200	200
8–10	—	—	—	—	10	100	1 000	1 000	600
6–8	—	—	—	100	1 000	5 000	5 000	5 000	2 500
4–6	—	—	1 000	10 000	30 000	30 000	25 000	15 000	5 000
2–4	—	10 000	100 000	200 000	200 000	100 000	50 000	25 000	10 000
0–2	1 000 000	1 000 000	800 000	600 000	400 000	200 000	100 000	50 000	20 000

(b) Nominal stress range induced by each size of wave (MPa)

	3–4	4–5	5–6	6–7	7–8	8–9	9–10	10–11	11–12
10–12	—	—	—	—	—	49	50	49	47
8–10	—	—	—	—	29	35	36	35	33
6–8	—	—	—	16	18	21	22	22	21
4–6	—	—	14	8	10	12	13	13	12
2–4	—	74	7	3	3	7	7	7	7
0–2	25	21	1	1	1	1	1	1	1

(c) Fatigue damage per year due to waves[a]

	3–4	4–5	5–6	6–7	7–8	8–9	9–10	10–11	11–12
10–12	—	—	—	—	—	0·000 0	0·000 0	0·000 1	0·000 1
8–10	—	—	—	—	0·000 0	0·000 0	0·000 2	0·000 2	0·000 1
6–8	—	—	—	0·000 0	0·000 0	0·000 1	0·000 1	0·000 1	0·000 0
4–6	—	—	0·000 0	0	0	0·000 0	0·000 0	0·000 0	0·000 0
2–4	—	0·008	0	0	0	0	0	0	0
0–2	0·012	0·003	0	0	0	0	0	0	0

[a] Values of less than 0·000 05 are shown rounded to 0·000 0.

$$D = 1/N = (1/2 \times 10^6)(\Delta\sigma/\Delta\sigma_{ref})^m \qquad (4)$$

$$= (1/2 \times 10^6)(\Delta\sigma/100)^{4\cdot38} = 8\cdot7 \times 10^{-16}\Delta\sigma^{4\cdot38}$$

where

$\Delta\sigma$ = hotspot stress range,
$m = 4\cdot38$,
$\Delta\sigma_{ref} = 100$ MPa.

When $\Delta\sigma = 147$ MPa

$$D = 2\cdot7 \times 10^{-6} \text{ per wave}$$

Part (a) of Table 1 indicates that the number of waves of 11 m $\times$ 10·5 s per year is 200 from all directions. However, only 25% come from directions which cause damage (see Fig. 4(c)). Hence, the total damage in 1 year due to waves of (10–12) m $\times$ (10–11) s is

$$D = 0\cdot25 \times 200 \times 2\cdot7 \times 10^{-6} = 0\cdot0001 \text{ per year}$$

This figure is indicated in the appropriate box in Part (c) of Table 1, which shows for all the other boxes with periods greater than 5 s the fatigue damage per year calculated in the same way. Values of less than 0·000 05 are shown rounded to 0·0000. The endurance limit in API[3] for S–N curve X is 35 MPa. When the hotspot stress range is below this the fatigue damage per year is shown as 0.

The columns in Part (b) of Table 1 containing the resonant periods of 4·7 and 3·6 s indicate the nominal stress ranges due to waves of the different heights at the resonant periods. Waves of 3 m $\times$ 4·7 s are estimated to induce a nominal stress range of 74 MPa, for which the hotspot stress range would be $\Delta\sigma = 74 \times 3 = 222$ MPa. The fatigue damage is

$$D = 8\cdot7 \times 10^{-16} \times 222^{4\cdot38} = 1\cdot6 \times 10^{-5} \text{ per wave}$$

The relevant box in Part (a) of Table 1 indicates that there could be 10 000 waves of (2–4) m $\times$ (4–5) s. Few of these waves are actually at the resonant period of 4·7 s. Ref. 4 showed that a reasonable estimate of the fatigue damage near the resonant peak could be obtained by considering all the waves within ±5% of the peak as inducing the peak stress range, and ignoring periods to each side. In this case ±5% of 4·7 s represents a range of 0·47 s, for which the number of waves of height (2–4) m is approximately 0·47 $\times$ 10 000 = 4700. However, less than 10% come from directions which cause damage (see Fig. 4(b)). Hence, the total damage in year due to waves of (2–4) m $\times$ (4–5) s is about

$$D = 0\cdot1 \times 0\cdot47 \times 10\,000 \times 1\cdot6 \times 10^{-5} = 0\cdot008 \text{ per year}$$

This figure is shown in Part (c) of Table 1. The figure in the box below was calculated in the same way.

The fatigue damage due to waves of (3–4) s was calculated in the same manner as that for those of (4–5) s except that 25% of waves were considered to come from critical directions (see Fig. 4(a)). It is evident that, due to the large number of these small waves, a significant amount of fatigue damage is calculated. A more rigorous analysis might break down the boxes with resonance into a number of smaller ranges of wave heights.

It is evident in Part (c) of Table 1 that the total fatigue damage due to all the waves is approximately $D = 0.025$ per year. API[3] recommends for fixed structures that D should not exceed 1·0 by the end of a design life of more than twice the service life (with greater factor of safety for critical elements whose sole failure could be catastrophic). Thus, a fixed structure with a service life of 20 years should not suffer damage at a rate faster than $D = 1/40 = 0.025$ per year. It is evident that the detail of this example would be borderline for a fixed installation and probably acceptable for a mobile unit if the exposure was no more severe than the case considered for most of the time. A small increase in leg height would lead to substantially more fatigue damage due to the increase in leg bending moment and increase of natural period. However, if the unit is likely to be on fatigue-vulnerable locations for only 30% (say) of its life, then stress levels could be increased by a factor of $0.3^{-1/4.38} = 1.3$.

2.6 Allowable stress approach

The detail analysed in Section 2·5 was selected to be borderline when analysed for fatigue so that an estimate could be made with the allowable stresses appropriate for the detail under large wave loading. It is calculated that when the independent leg unit example is subjected to a 23 m × 14 s wave (Stream Function) as a static load, without wind or current, the detail experiences a nominal stress of 140 MPa (20 ksi) and hotspot stress of 3 × 140 = 420 MPa (60 ksi). Consequently, the allowable peak hotspot stress for fatigue design for the selected location will be about 420 MPa (60 ksi) for the 23 m × 14 s wave. This stress can be compared with the allowable peak hotspot stress in API[3] under the fatigue design wave, which for 75 m water depth is about 540 MPa (78 ksi) for non-waterline members with 40 year design fatigue life. (See API[3] for specific values.) Thus, if an allowable stress approach were used for fatigue design of this jack-up with 40 year design fatigue life, the allowable stress could be less than the equivalent API figure for fixed structures. However, if the jack-up will spend only 6 years on fatigue vulnerable locations during its life, implying a 12 year design fatigue life, the allowable peak hotspot stress could be increased by a factor of 1·3 to 1·3 × 420 = 550 MPa (80 ksi), which is similar to API allowable stress for fixed structures.

2.7 Wind and current

Wind loading has been ignored in the above example because the wind load on a jack-up in 75 m water depth is generally smaller than the wave loading and because fluctuations of wind load do not have much energy at frequencies corresponding to the natural period of the structure. However, units designed for shallow water can be much more sensitive to wind loading than wave loading and it may be necessary to consider fatigue from turbulent wind loading as discussed in Ref. 5.

Currents can increase the fluctuations of loading from waves by a significant amount.[6] However, in the example above it was found that a current of 0·25 m/s (half knot) increased the stresses due to the smaller waves by only 3 or 4%, which is negligible in comparison with other approximations involved.

2.8 Spectral analysis

Considerable complications are involved in spectral analyses for fatigue damage of jack-ups if proper account is taken of the correlation between the sharp spikes of the structure's response to wave period and direction and the sharp spike of JONSWAP spectra. Significant fatigue damage can be calculated when the spikes interact, while less damage is done for most of the time when the spikes do not interact. The importance of coincidences of the spikes is illustrated in Fig. 5, as is explained in (a) to (e).

It is evident that sea-state (d), while being much smaller than sea-state (b), has caused a larger response because its frequency of maximum energy is coincident with the resonant frequency in (a). It is found that if T_z is increased or decreased by 10%, the lack of fit of the spikes in diagrams (a) and (b) results in a six fold reduction in fatigue damage. If numerical integration is to pick up the spikes with even moderate accuracy, it is necessary to use steps in frequency of the order of 0·05 Hz, i.e. about 0·1 s.

The sensitivity of fatigue damage to the fit of the spikes of the transfer function and the sea-state spectrum necessitates care being taken in estimating the appropriate occurrences of sea-states. Scatter diagrams typically have boxes of sea-state occurrences with steps in T_z of 1 s and

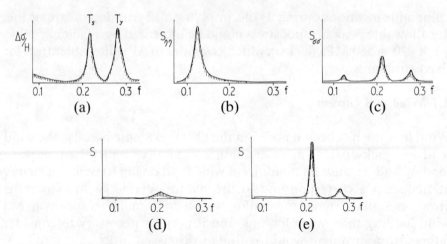

Fig. 5. Spectral analysis of independent leg unit example: (a) transfer function for stress range at the leg/spudcan connection of forward leg when subject to beam seas (with 2% damping) — the transfer function has spikes at the natural periods of $T_s = 4 \cdot 7$ s and $T_y = 3 \cdot 6$ s; (b) JONSWAP spectrum for a seastate of significant wave height $H_s = 4 \cdot 5$ m with zero crossing period $T_z = 6 \cdot 5$ s; (c) response spectrum of dynamic stress range derived from (a) and (b); (d) JONSWAP spectrum for sea-state $H_s = 1 \cdot 75$ m, $T_z = 3 \cdot 78$, drawn to same scale as (b); (e) response spectrum of dynamic stress range from (a) and (d).

steps in H_s of 0·5 m. Such diagrams need to be divided into smaller squares if a rigorous analysis is needed when the H_s and T_z are small. Many boxes on the scatter diagram are likely to contribute to the overall fatigue damage as was illustrated in Ref. 4. Spectral analyses have the potential of being more realistic than deterministic analyses by incorporating energy at the resonant frequencies in all sea-states. But it is not clear how accurate any spectrum is away from the central region, or how accurate transfer functions are for different sea-states. There is also the problem of the different sensitivities of moment range to wave direction at different frequencies (as in Fig. 4). To compound the problem, the stress range at each joint is sensitive to direction of bending.

Simplification is essential, and the following simplifying procedures are suggested for discussion:

(a) Use a static frame analysis of the leg to identify the stress concentrations most vulnerable to leg bending in each direction, and focus attention on critical joints and the associated directions of leg bending.

(b) Use a very simple three-mode dynamic frame analysis with single bars for legs and many load cases to determine the sensitivity of leg bending moments (in directions from (a)) to waves of different periods and directions.

(c) Split the spectral analysis into three parts.

(i) Quasi-static response to maximum energy region of spectra for large sea-states — the left blip in Fig. 5(c); this area of the response spectrum is proportional to the area under spectrum (b) times the square of corresponding ordinate in (a).

(ii) Dynamic response to low energy region of spectra for large sea-states — the right blips in Fig. 5(c); the areas of the response spectra are proportional to the ordinate of the spectrum in (b) and the areas under the squares of transfer function values in (a); directional response as in Fig. 4(a) and (b).

(iii) Dynamic response to high energy region of spectra for small sea-states with spike in (d) coincident with resonance in transfer function, as in Fig. 5(e); the area of response must be integrated from ordinate of spectra in (d) and square of ordinate of transfer function in (a).

3 TRANSIT CONDITIONS

3.1 Methods of calculation

Fatigue damage has occurred to the legs and jack-houses on a number of jack-ups during towage due to the cyclic loading from the legs during roll and pitch motions as shown in Fig. 6. Methods of calculation of fatigue damage for towage conditions were outlined in Ref. 7, and these methods are applicable to jack-ups. The structural mechanics of the legs under towage conditions are more easily predicted than the dynamic response of a unit in the elevated condition discussed above. However, fatigue damage calculations prior to a tow are still open to great uncertainty because of the difficulties in predicting the weather prior to the tow and in predicting the barge motions. In addition, fatigue damage on a tow is extraordinarily sensitive to the height of the legs above the guides. The stresses at the bottoms of the legs and supporting structure are approximately proportional to the cube of the leg height, while the fatigue damage is approximately proportional to the fourth power of stresses. Thus

$$\text{Fatigue damage} \propto (\text{Leg height})^{12}$$

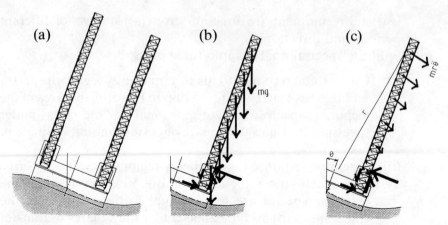

Fig. 6. Roll of jack-up on barge (a), gravity forces mg on legs (b), and inertia forces $mr\ddot\theta$ on legs (c).

A 20% reduction in leg height for a tow can reduce fatigue damage by about fifteen-fold. Calculations cannot accurately reproduce the uncertainties of weather, barge motions and fatigue calculations. Consequently, conservatism and simplicity are warranted.

The fatigue damage caused on a tow by the most severe storm is likely to be many times that due to the same period of moderate weather.[7] For a first approximation it can be assumed that all the damage is caused by the worst few hours or days of motion, while the majority of less severe motions are ignored. The problem for the forecaster is to predict the severity and duration of the worst storm, and the associated motions, for the chosen tow route.

3.2 Damage control criteria

The industry has been well served over the years by the criterion for ocean towage that structures should have adequate strength to resist extreme motions of 20° roll in 10 s period, when more precise motion predictions are not available. Additional factors may be added to account for shock loading (see Appendix). The following paragraphs show how the '20 in 10' criterion provides a basis for fatigue calculations in the absence of precise weather and barge motion predictions. It is also shown how fatigue considerations influence the allowable stresses appropriate for the extreme '20 in 10' motions.

It is assumed here that the severest storms on the tow cause half a day of motions with an extreme of '20 in 10'. Then assuming a Rayleigh distribution, the 20° extreme motion is equivalent to a significant motion of about $R_s = 20/2 \cdot 1 = 9 \cdot 5°$. According to Ref. 7 a significant motion of $9 \cdot 5°$ causes the same amount of fatigue damage as steady-state motion of about $R = 9 \cdot 5 \times 0 \cdot 85 = 8°$. Thus the fatigue damage due to 12 h of motions with an extreme of '20 in 10' can be calculated approximately from an equivalent steady-state motion of 8° in 10 s.

Reference 7 suggests for a once-in-a-lifetime tow of a jacket that the fatigue damage should not exceed $D = 0 \cdot 1$, because the remaining fatigue life is needed for operations. A similar argument applies for jack-ups, since significant fatigue damage can be caused to the bottoms of the legs and to guide structures during operations, as explained earlier in this paper. However, if a jack-up is likely to be subjected during its lifetime to a number of tows under severe environmental conditions, it may be appropriate to design for fatigue damage on a tow of only $D = 0 \cdot 05$. The chosen criteria of $D = 0 \cdot 1$ or $0 \cdot 05$ might have to be reduced for a particular rig in later life, if previous tows and operations cause significant fatigue damage.

In 12 h of motions there are about $n = 4300$ cycles of 10 s period. If the fatigue damage is to be limited to $D = 0 \cdot 1$, eqn (4)

$$D = 0 \cdot 1 = n/N = (4300/2 \times 10^6)(\Delta\sigma/100)^{4 \cdot 38}$$

and

$$\Delta\sigma = 100(0 \cdot 1 \times 2 \times 10^6/4300)^{1/4 \cdot 38} = 240 \text{ MPa}$$

Thus the hotspot stress range $\Delta\sigma$ in a vulnerable detail must be kept below 240 MPa under steady-state motions of 8° in 10 s. Under the extreme motion of '20 in 10', the hotspot stress range can be $(240 \times 20/8) = 600$ MPa (87 ksi). If it is assumed that the detail has a SCF = 3, then the maximum nominal stress range for '20 in 10' would be $600/3 = 200$ MPa (29 ksi).

The American Institute of Steel Construction (AISC)[8] has comparable provisions for fatigue of other structural details. The preceding analysis has assumed that 4300 cycles cause fatigue damage $D = 0 \cdot 10$. This is 20% of the allowable whole life damage, for which the equivalent number of cycles would be 22 000. ASIC Appendix B Loading Condition 1 relates to numbers of loading cycles between 20 000 and 100 000. It indicates that a Category D weld detail, for example, has an allowable stress range of 27 ksi (190 MPa), which is of the same order as the figure above.

If extreme motions on a tow are predicted to exceed '20 in 10', or if the duration of extreme loading is likely to exceed 12 h, then lower allowable stresses are relevant. Appropriate values can be calculated by following the above calculations using the predicted extreme motions and storm duration.

If fatigue damage is not considered, a unit might be accepted for towage in accordance with the '20 in 10' criteria with maximum nominal stresses of the order of 350 MPa (50 ksi). If it is assumed that the maximum reversal of stress during roll motions is -70% of the maximum, the overall nominal stress range would be about (1·7 $\times$ 350) = 600 MPa (85 ksi). In order to meet the above fatigue criterion, the legs of the unit would have to be shortened to reduce the nominal stress range to about 200 MPa, as indicated in the preceding calculation. Since stresses near the leg bottoms are approximately proportional to leg height to power of 3, the reduced length would need to be $(200/600)^{1/3} =$ 0·69, i.e. one third less. The fatigue damage without leg shortening would be about $(600/200)^{4·38} = 100$ times the limit recommended above.

3.3 Shock loading

The Appendix explains how shock loading from the legs rocking in their guides increases the moment at the guides. In 20° rolls the increase in moment is relatively small because any rocking rotation is small compared to 20°. However, with smaller motions, such as 8°, the shock loading is proportionally much greater. In the example in the Appendix, shock loading increases the leg and guide moment by about 20% in 20° roll motions, and by about 40% in 8° motions. An increase of 40% in stresses would lead to an increase in fatigue damage of $(1·4)^{4·38} =$ four-fold, if shock oscillations occurred on every roll motion. Hence, it is most important that legs are secured properly, otherwise the legs should be shortened further.

4 CONCLUSIONS

This paper has shown that fatigue damage calculations for a jack-up in the operating condition are extremely sensitive to

1. resonances of the unit at its natural periods in surge, sway and yaw (more fatigue damage is calculated at resonances in small waves than in large waves);
2. directions of the waves, particularly those causing resonance (a structural detail may only suffer significant fatigue damage at resonance in waves from 10% of directions);
3. interaction of leg positions with wavelengths at resonance;
4. derivation of wave data from observations (most fatigue damage is

calculated for wave sizes and sea-states at the edges of observed distributions; the data gathering for such waves may be inaccurate).

It has been shown that the API allowable peak hotspot stresses provide an indication of the allowable values that could be used for fatigue design of a jack-up which will work on fatigue vulnerable locations for part of its life.

It has been shown for a jack-up under transit conditions that

1. Fatigue damage calculations for roll and pitch motions are dominated by the few hours or days of worst motions; damage due to the other days may be negligible in comparison.
2. It is suggested that calculated fatigue damage in details should be kept below about $D = 0.1$ per tow, in order that whole life damage should remain below $D = 0.5$; lower limits may be appropriate for units that are likely to be (or have already been) exposed to severe motions on several tows.
3. If the industry standard for transit motions of 20° roll with 10 s period is assumed to represent the extreme motion during 12 h or severest weather, then in order to limit fatigue damage the allowable peak hotspot stress range under the '20 in 10' motion should be about 600 MPa (87 ksi); the equivalent allowable nominal stress range (assuming a stress concentration factor of 3) is about 200 MPa (29 ksi).
4. If fatigue is ignored and a unit is checked for '20 in 10' motions with nominal stresses of the order of 350 MPa (50 ksi) then a 12 h storm with '20 in 10' extreme motions is likely to cause about 100 times more fatigue damage than the limit suggested above; removal of the top third of legs would avoid this excessive damage.
5. Fatigue damage during transit at the bottoms of the legs and supporting structures is approximately proportional to the leg height to power of 12.
6. Legs must be secured properly in order to avoid large increases in fatigue damage due to shock loading of legs rocking in their guides.

REFERENCES AND BIBLIOGRAPHY

1. Brekke, J. N., Murff, J. D., Campbell, R. B. & Lamb, W. C., Calibration of jack-up leg foundation model using full-scale structural measurements (OTC 6127). Paper presented at 21st Offshore Tech. Conf., Houston, 1988.
2. Hambly, E. C., Oil rigs dance to Newton's tune. *Proc. Royal Instn,* **57** (1985) 79–104.
3. American Petroleum Institute, *Recommended Practice for Planning Designing*

and Constructing Fixed Offshore Platforms, (17th edn). API, Washington DC, USA, 1987.

 4. (a) Hambly, E. C., Fatigue vulnerability of jack-up platforms. *Proc. Instn. Civil Engrs (Part 1),* **78** (1985) 161–78. (b) Discussion by Carlsen, C. A., Dolan, J. L., Ellis, L. G. & Hambly, E. C. *Proc. Instn. Civil Engrs, (Part 1),* **80** (1986) 291–6.
 5. Davenport, A. G. & Hambly, E. C., Turbulent wind loading and dynamic response of jack-up platforms. Paper presented at 16th Offshore Tech. Conf., Houston, 1984.
 6. Ahilan, R. V., An approach to the fatigue analysis of jack-up legs. In *Mobile Offshore Structures*, ed. L. F. Boswell, C. A. D. Mello & A. J. Edwards. Elsevier Applied Science, London, 1988, 104–24.
 7. Hambly, E. C., Edwards, A. J., Kohli, C. & Miller, B. L., Fatigue consideration for ocean towage (OTC 4163). Paper presented at 13th Offshore Tech. Conf., Houston, 1981.
 8. American Institute of Steel Construction, *Specification for the Design, Fabrication and Erection of Structural Steel for Buildings,* latest edition. AISC, Chicago, USA.
 9. Department of Energy, *Offshore. Installations: Guidance on Design and Construction,* HMSO, London.
10. Department of Energy, *Background to New Fatigue Design Guidance for Steel Welded Joints in Offshore Structures.* HMSO, London, 1984.
11. Det Norske Veritas, Fatigue strength analysis for Mobile Offshore Units, Classification Note No. 30·2, 1984, August, 1–60.
12. Det Norske Veritas, Strength analysis of main structure for self elevating units Classification Note No. 31·5, 1984, May, 1–94.

APPENDIX

SHOCK LOADING OF LEGS ROCKING IN GUIDES

Figure A1 illustrates the roll of a barge with a leg loose in the guides. The barge rolls with single amplitude R and period T; the leg rocks through angle s to each side of centre of guides. Sketches A, B, C, D show the barge at four positions of roll — at extreme roll to port in A, level in B, at point where the leg first hits the opposite side of the guide in C, and at extreme roll to starboard in D. The graph shows by the continuous line the change with time of roll angle θ of the barge and guides, and by the dashed line the angle $(\theta + \phi)$ of the bottom of the leg, where ϕ is the angle between the leg and the guides.

The barge roll velocity is the gradient of the curve of roll in Fig. A1(a): it is zero as the roll reverses at A, it is maximum at B, and zero at D as it reverses again. As the barge passes B the barge roll velocity starts to slow down, but the leg is not restrained and continues along bc; it continues along bc until it has overshot the angle of the guides by s at cC. The barge roll angle and roll velocity are given by

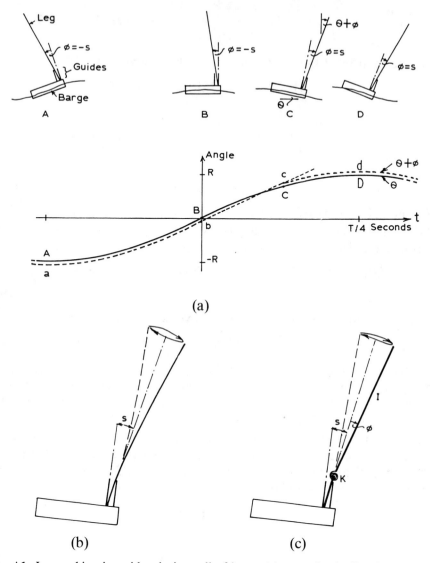

Fig. A1. Leg rocking in guides during roll of barge (a), post-shock vibration (b), and equivalent spring system (c).

$$\theta = R\sin(wt) \qquad d\theta/dt = Rw\cos(wt) \qquad \text{(A1)}$$

where $w = 2\pi/T$

For small values of t these can be approximated by

$$\theta = R((wt) - (wt)^3/6) \qquad d\theta/dt = Rw(1 - (wt)^2/2) \qquad \text{(A2)}$$

At $t = 0$

$$(d\theta/dt)_0 = Rw$$

The inclination of the leg $(\theta + \phi)$ along bc is given by

$$(\theta + \phi) = -s + (d\theta/dt)_0 t = -s + Rwt \tag{A3}$$

At cC, $t = t_c$ and $\phi = s$. Hence

$$s = \phi_c = (\theta + \phi)_c - \theta = -s + Rwt_c - R((wt_c) - (wt_c)^3/6) \tag{A4}$$

giving

$$(wt_c)^3 = 12s/R \tag{A5}$$

The shock loading on the leg at stage C results from the sudden stopping of the leg bottom relative to the guides. From eqns (A2) and (A3), $(d\phi/dt)$ during stage BC is obtained:

$$(d\phi/dt) = Rw - d\theta/dt = Rw - Rw(1 - (wt)^2/2) = Rw(wt)^2/2 \tag{A6}$$

which with eqn (A5) gives the relative velocity at C of

$$(d\phi/dt)_c = (Rw/2)(12s/R)^{2/3} = (R\pi/T)(12s/R)^{2/3} \tag{A7}$$

When the bottom of the legs are suddenly forced to move with the guides the upper regions of the leg overshoot and bend under the inertia forces, as shown in Fig. A1(b), and continue to vibrate at the natural period of the elevated legs. This behaviour can be analysed with the simplified model in (c), in which the flexible leg is represented by a stiff leg, with the same moment of inertia I about the bottom, and with a spring of stiffness K which gives the same natural period T_n, for the leg. The leg vibration with maximum angle r is given by

$$\psi = r\sin(2\pi t/T_n)$$

where

$$T_n = 2\pi(I/K)^{1/2} \tag{A8}$$

time t is now measured from instant of impact in guides

$$d\psi/dt = r(2\pi/T_n)\cos(2\pi t/T_n)$$

At impact the leg angular velocity $(d\psi/dt)$ relative to guides is equal to the sudden stopping of $(d\phi/dt)$ of the bottom of the legs, in eqn (A7).
    Hence

$$(d\psi/dt)_c = r2\pi/T_n = (R\pi/T)(12s/R)^{2/3} \tag{A9}$$

The moment in the leg due to the shock loading is

$$M = Kr = (KRT_n/2T)(12s/R)^{2/3} \qquad \text{(A10)}$$

Assume that the elevated legs of an independent leg unit above the upper guides have $I = 4 \times 10^5 \, t\text{m}^2$ and $K = 5 \times 10^6$ kN m/radian giving $T_n = 1\cdot8$ s. Assume also a clearance of $0\cdot02$ m with height between guides of 20 m, giving $s = 0\cdot02/20 = 0\cdot001$ radians. For a roll with $R = 20° = 0\cdot35$ radians and $T = 10$ s, eqn (A10) gives $M = 17$ MN m, which is about 15% of the primary roll moment of 115 MN m due to '20 in 10'. While with $R = 8° = 0\cdot14$ radian eqn (A10) gives $M = 12$ MN m which is 26% of the primary moment of 46 MN m due to 8° roll in 10 s. If this leg had been three times as stiff with $K = 1\cdot5 \times 10^7$ kN m/radian, T_n would have been 1 s, and the shock loading moments would have been 25% of primary moment in 20° roll in 10 s, and 45% of primary moment in 8° roll in 10 s.

The Weldability of Steels Used in Jack-Up Drilling Platforms

R. J. Pargeter

The Welding Institute, Abington Hall, Abington, Cambridge CB1 6AL, UK

ABSTRACT

Weldability problems in jack-up rigs arise principally in leg construction where particularly high-strength steels (~690 N mm^{-2} yield) are used for chords and racks. Attainment of mechanical properties is not usually difficult, although procedural trials are advisable.

Of potential fabrication problems, hydrogen cracking (heat-affected zones or weld metal) is of greatest concern, and consideration of solidification cracking is advisable. Lamellar tearing is not generally a problem with modern steels, and the risk of stress-relief cracking will be confined to the limited number of joints which can be heat-treated.

Fatigue cracking is probably the major cause of service failure of jack-up rigs, and the use of high-strength steels, which permits higher static stress limits, can exacerbate this problem. Hydrogen-induced stress corrosion can also occur, either due to cathodic polarisation or corrosion, particularly in the presence of H_2S in foul sea-water. For this reason weld hardness limits need to be maintained.

Key words: jack-up rigs, weldability, high-strength steel, fabrication cracking, service cracking, mechanical properties.

1 INTRODUCTION

From a structural and fabrication standpoint, jack-up rigs are clearly divided into two parts — the hull and the legs. The hull is essentially a plate structure, built using largely conventional shipbuilding materials and practices. The only area here which might give a shipbuilder cause for concern is the very heavy construction around the leg wells. By contrast with the hull, the legs are generally of tubular construction, and use steel significantly stronger than that used in most other shipbuilding or offshore construction. Furthermore, dimensional tolerances on the legs are very strict, and it is in the fabrication of these that special problems are posed by jack-up rigs. This paper, therefore, concentrates on the weldability of leg materials.

Weldability is a term with many definitions, and another will not be added here. It is, however, necessary to realise that the weldability of a material is concerned with the soundness, properties, and serviceability of welded joints which can be made in that material. Having reviewed the materials which have commonly been used in the legs of jack-up rigs, this paper, therefore, looks at the attainment of required mechanical properties, fabrication problems, and weld related service problems, before briefly describing the choices of welding process and consumables which are open to the designer.

2 MATERIALS

Jack-up rig leg construction may vary in detail, but will generally consist of chords, braces and racks of different materials. Both chords and racks use steel with around 690 N mm^{-2} yield strength and 800 N mm^{-2} tensile strength. These are often referred to as '80 grade' steels (80 kgf/mm^2 tensile strength), but should not be confused with, for example API 5L X80, which is 80 ksi (552 N mm^{-2}) minimum yield. Tubular steel of around 1 m diameter and 35 mm wall thickness is usually used for chord members, and the racks are cut from plates, typically 125–150 mm thick. A much wider range of strength levels is used for the braces, ranging from a similar strength to that used for chords and racks down to API 5L grade X65 (65 ksi, 448 N mm^{-2} yield strength) for the major braces.

To obtain 690 N mm^{-2} yield strength, quenched and tempered steel is necessary. Tubular chord members are too large to be made in useful lengths as seamless pipe, and the combination of strength and thickness puts them outside the capacity of conventional UOE seam welded pipe. One solution to this has been to form the pipe prior to heat treatment and to select welding consumables such that the completed pipe can be quenched and tempered.[1] For braces, seamless pipe is generally favoured.

High strength is necessary both to keep section sizes convenient and to limit weight. This is particularly important in legs, as they may otherwise impair the stability of the rig in transit. In addition, however, for operation in cold environments, a toughness requirement is imposed, typically a minimum Charpy energy of about 35 or 40 J at $-40\,^\circ$C for North Sea service.

The chemical composition of rolled 690 N mm^{-2} tensile-strength steels is normally based on a Cu–Ni–Cr–Mo–V–Al–B alloy route. Carbon levels are usually just over 0·1% and manganese just below 1%. Nickel levels tend to be higher, at the expense particularly of carbon and vanadium, in higher toughness grades, and sulphur is kept as low as possible. Examples of chemical compositions of such grades of steel are given in Table 1. Centrifugally cast pipes, which are particularly suitable for large diameter thick wall forms, have also been proposed, and the specified composition of up to 800 mm diameter, 40 mm wall-thickness pipe incorporated into a French design[2, 3] are included in Table 1. Lower strength seamless pipes for braces would be of lower alloy level and, in general, be of less concern with regard to weldability, both because of the composition and the thinner section sizes.

3 MECHANICAL PROPERTIES OF WELDED JOINTS

Welded joints in jack-up rig legs need to meet strength and toughness requirements in weld metal and heat-affected zones (HAZ). Welding consumables of adequate strength are readily obtainable for submerged-arc, gas-shielded and manual metal-arc processes, typically using a Ni–Cr–Mo alloy route. Matching strength consumables such as AWS E11016-G MMA electrodes or AWS F11A6-EG-G submerged-arc consumables are necessary for butt joints between two 690 N mm^{-2} yield materials (e.g. chord girth welds), but wherever possible lower strength consumables should be considered (e.g. chord: lower strength brace welds) in order to reduce problems such as weld metal hydrogen cracking as discussed below. Heat affected zone strength may be reduced to below plate levels in the lower temperature regions, particularly of high heat input welds, where the plate is effectively over tempered. This is not usually a practical problem with typical arc energies, however, and in thick plates, submerged arc welds of up to 3 or 4 kJ/mm are normally possible.

There is only a slight effect of postweld heat treatment (PWHT) on strength, although it is necessary to ensure that this is carried out at below

TABLE 1

Typical Chemical Composition of 690 N mm⁻² Yield Strength, 800 N mm⁻² Tensile-Strength Steel for Chords and Racks

	Rolled plate/pipe							Centrifugally cast pipe	
	Ref. 4	*Ref. 6*	*Ref. 7*	*Ref. 16*	*Ref. 24*	*Ref. 25*	*Ref. 26*	*Ref. 27*	*Ref. 2,3*
C	0·1	0·12	0·12	0·12	0·11	<0·20	<0·18	~0·08	<0·010
Si	0·26	0·25	0·24	0·22	0·26	0·15	<0·500 –0·70	~0·33	<0·40
Mn	0·98	0·87	0·90	0·86	0·85	<1·7	<1·5	~2·2	1·7 –2·2
P	0·014	0·013	0·009	0·013	0·007	<0·030	<0·020	~0·01	<0·020
S	0·002	0·005	0·005	0·004	0·001	<0·010	<0·010	~0·01	<0·012
Cu	0·25	0·18	0·25	0·23	0·50			~0·01	
Ni	0·54	0·79	0·90	1·38	1·30		<1·5	~3·0	2·50 –3·50
Cr	0·48	0·62	0·58	0·72			<1·2	0·04	
Mo	0·38	0·38	0·45	0·38	0·45	<0·07	<0·5	~0·25	
V		0·044	0·04	0·04	0·03		<0·08		
Al		0·095	0·051	0·051			0·015 –0·050	~0·04	
Ti									
B			0·0014	0·0013	0·002	<0·005			
N			0·0037					~0·02	
Nb								~0·07	0·05 –0·08
Carbon equivalent	0·49	0·54	0·56	0·60	0·47			~0·7	

the original tempering temperature of the plate to avoid a general loss of strength. In contrast, toughness levels both in the heat affected zone and weld metal, can be severely reduced by PWHT,[4] and different welding consumables may be selected depending on whether the joint is to remain as-welded or be PWHT. It is difficult to predict toughness levels in the as-welded or PWHT condition, however, and procedural trials are necessary for complete confidence. In general the lowest toughness region in rack-rack or rack-chord welds has been found to be the weld metal, followed by the fusion line,[5-7] but in all these cases, Charpy energies of better than 40 J at $-40\,^{\circ}$C were obtained in welds of between 1·7 and 4·1 kJ/mm.

4 POTENTIAL FABRICATION PROBLEMS

Material related fabrication problems may include lamellar tearing, hydrogen cracking, solidification cracking and stress relief cracking. Lamellar tearing may be discounted for all practical purposes as although there are many highly restrained joints, modern clean materials have a very high resistance to lamellar tearing. As stated earlier, sulphur levels in the rolled 690 N mm^{-2} yield materials are typically 0·001–0·005%, and short transverse reduction of area (STRA) values of up to 40% have been reported.[1] To put this in context, about 25% STRA is generally considered to provide almost complete immunity from lamellar tearing in common highly restrained joints[8-10] and values of <20% STRA are unlikely with sulphur contents of <0·005%. In cast materials, even higher STRA levels may be attained despite higher sulphur, owing to the better (more globular) inclusion morphology although this gives no further benefit with regard to lamellar tearing.

Hydrogen cracking is probably the major concern when welding high-strength steels, and needs to be guarded against both in the HAZ and weld metal. So far as HAZ cracking is concerned, guidance can be obtained from BS5135 (Ref. 8) although typical carbon equivalent (CE) levels of the 690 N mm^{-2} yield steels (0·5–0·6) are near or above the limits of applicability of this document (0·54 maximum) (The Welding Institute book,[11] on which Appendix C of BS5135 is based, does refer to higher CE steels). Alternatively, AWS D1.1 appendix Xl, which normally goes up to a CE value of about 0·7, gives guidance based on hydrogen content, restraint, and P_{cm} value,[†] or the approach proposed by Yurioka

---

$$^{\dagger}P_{cm} = C + \frac{Si}{20} + \frac{Mn}{20} + \frac{Cu}{20} + \frac{Ni}{60} + \frac{Cr}{20} + \frac{Mo}{15} + \frac{V}{10} + 5B$$

et al.[12] may be followed. This involves the determination of a cracking index, based on composition, hydrogen content, and restraint, and prediction of preheat through the critical time to cool to 100 °C, derived from the cracking index. The work on which this was based included WES HW 70 steel of composition similar to those given in Table 1. The important point to realise is that these steels are hardenable, and precautions are necessary, particularly in highly restrained joints.

Less guidance is available on the avoidance of weld-metal hydrogen cracking, but, as with HAZ cracking, the risk increases with alloy level (in this case in the weld metal) and with joint restraint, and can be reduced by maintaining low hydrogen levels. In the roots, particularly of single-sided welds, a number of factors combine to increase the risk of weld-metal cracking, and slightly lower-strength consumables are often used for this run. It is, for example, often difficult to ensure that the joint preparation is completely clean and dry in the root; also, the root gap provides a stress concentration for initiating cracking; and furthermore, preheat (or interpass) levels will be at their lowest for this run. From the small-scale test data generated by Hart,[13] it is apparent that, at low hydrogen levels, tougher weld metals with finer microstructures and Ni additions have greater resistance to cracking. When welding steels of unequal strength (e.g., chord–brace welds) the risk of weld-metal hydrogen cracking can be minimised by matching the lower-strength material.

Two factors which affect the risk of solidification cracking are restraint and weld-metal composition. Restraint tends to be high in jack-up rig legs, which increases the likelihood of this type of cracking. The effect of composition is described by the UCS formula:[8, 14]

$$UCS = 230C^* + 190S + 75P + 45Nb - 12\cdot3Si - 5\cdot4Mn - 1$$

where C^* represents carbon content unless it is <0·08%, when a value of 0·08 shall be used. A low risk of cracking is generally expected at UCS values below about 10, and a high risk above 30.[8] This formula, however, is only strictly applicable for weld-metal nickel contents of <1%, chromium contents of <0·47%, and molybdenum contents of <0·4%, and it is known that, although chromium and molybdenum may as ferrite formers be beneficial, nickel increases the risk of cracking.[15] It should be remembered that weld metal is composed of both welding consumables and plate, and in high dilution regions it may be necessary to modify consumable composition to reduce the risk of cracking.[1, 7]

If the joints are to be post-weld heat-treated (PWHT) it must be recognised that the 690 N mm^{-2} yield steels can be susceptible to stress-relief (or reheat) cracking. For this reason, steel makers try to maintain low levels of carbide formers, and particularly of vanadium,[1] and impurities.[16] This form of cracking occurs during PWHT owing to a lack of creep ductility, generally caused by a strengthening within the grains from precipitation hardening (hence, the importance of carbide formers) and grain-boundary weakening by impurities. Thus, if a differential in grain interior–boundary strength develops during PWHT before residual stresses have been relaxed by creep, grain-boundary cracking may occur. A review of reheat cracking was produced by Dhooge and Vinkier in 1986.[17] Weld regions are susceptible to stress-relief cracking not only because of the overall levels of residual stress but also because of the stress concentrations at weld toes and the coarse grain size in heat-affected zones. Some reduction in the risk of cracking can, therefore, be obtained by blending out stress concentrations in critical areas prior to PWHT, and by using multipass procedures which give as much grain refinement as possible in the HAZ.[17]

5 POTENTIAL WELD-RATED SERVICE PROBLEMS

Excluding damage to jack-up rigs from accidents such as punch-through of the sea bed, the major cause of failure of jack-up rigs is fatigue cracking in the legs. This may occur either during service, or during towing when there are large unsupported free lengths of leg extending above the hull. Fatigue cracks almost invariably initiate at welds in such structures because of the geometrical stress concentrations and residual stress. Although the latter can be alleviated to some extent by PWHT, it is not possible to stress-relieve all welds in this way (only prefabricated parts of the leg can be heat-treated), and long-range stresses from connection to other parts of the leg cannot be avoided. It is unlikely that any real benefit to fatigue life would be conferred by PWHT. Furthermore, dimensional tolerances on legs are quite tight, so that PWHT may not be desirable because of the risk of distortion. The risk of stress-relief cracking has been mentioned in the previous section.

Although some improvement in the fatigue strength of a welded joint can be achieved through dressing the weld toes, this is generally impractical for a complete structure. If stresses cannot be kept down to acceptable levels in node regions to provide adequate fatigue life for a welded joint, some benefit can be gained from using forged or cast nodes,[18, 19] thereby moving the welds to less-highly-stressed regions. One reason why jack-up rig legs are particularly susceptible to fatigue cracking is the use of high-strength materials to save weight.[20] This allows the designer to work with higher static stress levels, but, unfortunately, higher-strength steels do not have higher fatigue strengths at welded joints. This is because, owing to the presence of small intrusions at weld toes, there is no initiation stage in fatigue crack growth at welds, and crack propagation rates are not significantly affected by steel strength.

In any hardenable welded steel operating in sea-water, there is a potential risk of hydrogen-induced stress corrosion in hard HAZs. The hydrogen may come from corrosion or cathodic protection, and the amount of hydrogen pick-up may be increased in the presence of hydrogen sulphide. This can be present particularly in the muds, and stagnant water inside parts of structures such as spud-cans, as a result of the action of sulphate-reducing bacteria. Welded regions are particularly at risk because of the combination of local hard regions and high residual stresses. In clean sea-water, with normal levels of cathodic protection, it is probably acceptable to have hardness levels of up to around 400 HV in grade-50 steel HAZs. However, if excessive protection arises from, for example, a malfunction of an impressed current system, or there is hydrogen sulphide present, cracking may occur at much lower hardness levels. Furthermore, cracking is dependent on applied and/or residual stress, as well as material susceptibility (crudely described by hardness). Thus, in higher-strength steels where higher stresses can be developed, it may even be that hardness levels which can be tolerated are lower than in lower-strength steels. At present the detailed relationships between environment and hardness are not clearly established, but it is unlikely that environments sufficiently severe to cause cracking at 350 HV maximum will occur in normal service. The lower bound can be taken as 250 HV for acid sour service such as can be experienced in a sour-gas pipeline, but this is not relevant to sea-water environments for which it would be excessively conservative.

6 CHOICE OF WELDING PROCESS AND CONSUMABLES

The selection of a welding process is governed principally by the geometry of the joint, and by economic factors. For long straight welds in thick material, such as rack–chord joints, high-productivity mechanised processes are favoured, and narrow-gap joint preparations help both with productivity and control of distortion.[21-23] Dimensional tolerances on jack-up rig legs are particularly tight, and distortion has to be a major consideration in the detailed design of any welding procedure[4,7] including, for example, balanced welding and careful control of local preheating. For node joints between tubular components, the continually varying preparation around the joint means that manual processes have to be used.[7] For this, manual-metal arc welding is still the most common, although flux-cored arc welding must be a strong contender. Submerged-arc is possibly the most popular mechanised process for long straight joints, although various MIG variants have been used, particularly in Japan.[22, 23] The principal factors involved in the selection of particular consumable compositions are mechanical properties and resistance to cracking, and these have been covered in previous sections.

REFERENCES

1. Taira, T., Hirabayashi, K., Ume, K., Ishihara, T., Yamada, M. & Watanabe, T., Properties of high strength UOE pipes for jack-up rig. Technical Report (Overseas) No. 33, Nippon Kokan, December 1981, pp. 6–16.
2. Royer, A., The horizontal centrifugal spinning, a technique for the manufacture of large diameter heavy wall pipes. International Colloquium on the metallurgy of heavy section parts, Mons, France, 12–13 May 1981.
3. Royer, A., Dumas, B., Metauer, G. & Toll-Duchanoy, T., Centrishore centrifugally cast pipe elements for deep sea utilisation. Deep Offshore Technology Conference, Palma, Spain, October 1981.
4. Akahide, K., Kikukawa, S., Hashimoto, O. & Shimizu, T., Recent advances in the welding technology on the fabrication of jack-up drilling rig (Paper OTC4597). In *Proceedings of the Offshore Technology Conference (Vol. 3)*, 1983, pp. 145–52.
5. Okano, S., Yano, K., Kaji, H. & Takisawa, K., Development of heavy thick HT80 steel plates for racks of jack-up rigs. *Transactions of the Iron and Steel Institute of Japan*, **27**(1) (1987) B14.
6. Hirano, Y., Narumoto, A., Kawasaki, H. & Koyamo, Y., The toughness and fatigue strength of welded joint of 800 N/mm² class seamless steel pipe with good weldability for jack-up rig. In *Proceedings of the Offshore Mechanics/Arctic Engineering/Deepsea Systems (Vol. 1)*, ASME, 1982, pp. 79–83.
7. Hamada, K., Wake, T., Koyama, Y., Nakatsuji, K. & Kusuhara, Y., Making and fabricating of steel components for jack-up (offshore) rig legs. Technical Report No. 6, Kawasaki Steel, September 1982, pp. 80–97.
8. British Standards Institution, Process of welding carbon and carbon manganese steels (BS5135), BSI, 1984.
9. Australian Welding Association, Control of lamellar tearing. Technical Note No. 6, AWRA, November 1985.
10. Farrar, J. C. M., Ginn, B. J. & Dolby, R. E., The use of small scale destructive tests to assess susceptibility to lamellar tearing. International Conference on Welding in Offshore Constructions, Newcastle, 26–28 February 1974, Paper 8, pp. 98–113, the Welding Institute.
11. Coe, F. R., *Welding Steels without Hydrogen Cracking*. The Welding Institute, Cambridge, UK, 1973.
12. Yurioka, N., Suzuki, H., Ohshita, S. & Saita, S., Determination of necessary preheating temperature in sled welding. *Welding Journal* (research supplement), **62**(6) (1983) 147s–53s.
13. Hart, P. H. M., Resistance to hydrogen cracking in steel weld metals. *Welding Journal* (research supplement), **65**(1) (1986) 14s–22s.

14. Bailey, N. & Jones, S. B., *Solidification Cracking of Ferritic Steels During Submerged-Arc Welding.* The Welding Institute, Cambridge, UK, 1977.
15. Marshall, W. K. B., A note on the relationship between chemical composition and hot cracking in mild and alloy steels. *British Welding Journal,* **7**(7) (1960) 451–3.
16. Iwasaki, N., Tagawa, H., Watanabe, T., Yamada, M., Nagamine, T. & Endo, G., Properties of class 80 kgf/sq mm. [785 MPa] grade high strength heavy section steel plates for jack-up rig racks. Technical Report (Overseas) No. 30, Nippon Kokan, December 1980, pp. 1–12.
17. Dhooge, A. & Vinkier, A., Reheat cracking — a review of recent studies. *International Journal of Pressure Vessels and Piping,* **27**(4) (1987) 239–69. Also in *Welding in the World,* **24**(5–6) (1986) 104–27.
18. Nakajima, T., Konomi, M., Kunitomi, A., Yano, T., Nagashima, A. & Asano, K., Full size test on leg node of jack-up rig. Technical Report (Overseas) No. 25, Nippon Kokan, September 1978, pp. 65–78.
19. Hori, K., Tanaka, T., Muka, T. & Watanabe, S., The development of high tensile casting steel leg nodes for jack-up type rigs. *Sumitomo Search,* **17** (1977) 122–8.
20. Hambly, E. C., Fatigue vulnerability of jack-up platforms. In *Proceedings of the Institute of Civil Engineers (Part 1: Design and Construction) (Vol. 77),* 1985, pp. 161–78.
21. Sugioka, I. & Sueda, A., Submerged arc narrow gap welding process 'SUBNAP'. In *Narrow Gap Welding (NGW), The State-of-the-Art in Japan.* Japan Welding Society, 1986, pp. 177–81.
22. Nakajima, H., Nagai, A. & Minehisa, S., Rotating arc narrow gap MIG welding process. In *Narrow Gap Welding (NGW), The State-of-the-Art in Japan.* Japan Welding Society, 1986, pp. 65–73.
23. Okuda, N., Kashimura, T. & Saita, H., Narrow gap GMA welding process twist arc welding process. In *Narrow Gap Welding (NGW), The State-of-the-Art in Japan.* Japan Welding Society, 1986, pp. 46–55.
24. Creusot Loire, SE 702, steelmaker's literature.
25. Svenkst Stal, OX 812, steelmaker's literature.
26. Fabrique de Fer de Charleroi, Supraslim E690, steelmaker's literature.
27. Pont á mousson, centrishore V, manufacturer's literature.

Collision Damage of Jack-Ups

Charles P. Ellinas, Raymond Kwok & Kevin A. J. Williams

Advanced Mechanics and Engineering Ltd, 4 Frederick Sanger Road,
Surrey Research Park, Guildford, Surrey, GU2 5YJ

ABSTRACT

North Sea collision accident records covering a period of ten years indicate that the risk of collisions involving jack-ups is not dissimilar to that for semi-submersibles or fixed jacket structures. However, jack-ups are much more flexible than jackets and have a much lower degree of redundancy. Their response to collisions and their ability to absorb impact energy is, therefore, expected to be considerably different in comparison to jackets. This paper examines available data and information regarding the capability of jack-ups to withstand collision impacts and investigates the level of local damage that can potentially be caused to jack-up legs due to accidental collisions.

Key words: jack-up, accidental impact, damage energy absorption, integrity.

1 INTRODUCTION

One of the major risks to offshore installations in the North Sea is impact and damage caused through collisions, mostly with attendant vessels. This was the subject of a major study initiated by the Department of Energy, which ended with the publication of the Department's revised, draft, guidance notes for the protection of offshore installations against impact[1] and a supporting background report.[2] The study examined all types of offshore installations, fixed structures, semi-submersibles and jack-ups, and covered the period between 1975 and 1986. In total about 240 collision accidents were analysed, drawing the following major conclusions:

- most incidents are caused by supply vessels, but no clear trends exist with time, especially with regards to the rate of serious accidents;
- the most common operating circumstances at the time of collisions are installation approach and bulk transfer, while misjudgement and equipment failure are the most frequent causes of incidents; weather does not appear to be a dominant cause of serious accidents;
- about 90% of all reported collision accidents involved less than 0·5 MJ of energy absorbed by the installation.

However, the study also concluded that as jack-ups are comparatively more flexible than fixed jackets, during collisions they would absorb a comparatively smaller proportion of impact energy, through the development of plastic deformations and permanent damage, than fixed steel jacket platforms.

Guidance for the design of jack-ups to ensure adequate resistance against collisions is provided by Det norske Veritas (DnV).[3] In analysis performed on jack-ups typical of those operating in the North Sea,[4,5] DnV concluded that significant levels of impact energy can be absorbed by jack-up legs without developing serious levels of damage. However, the results of fatigue analysis of jack-ups[6] suggest that even though damage may not be critical from the point of view of strength, fatigue considerations may require early repair.

Available information and data on collisions and damage of jack-ups are reviewed in this paper. Following analysis of a typical structure subject to collision impact, the resulting levels of local damage are assessed and their implications for jack-up integrity are investigated.

2 NORTH SEA EXPERIENCE

The operating experience of jack-ups in the North Sea, measured in rig-years, has been increasing steadily since the late seventies. Even though their area of operation is limited to the southern North Sea, about a third of mobile offshore rig activity in the UK sector of the North Sea now involves jack-ups. This is illustrated in Fig. 1. This increased activity has led to a somewhat larger number of collision accidents per year, but in general, in common with other types of offshore structures, the accident rate per rig-year appears to be independent of time.

As shown in Table 1, after semi-submersibles, jack-ups have the highest collision accident rate, with about one accident per rig every four years.

However, the risk of a serious collision, defined as a collision which

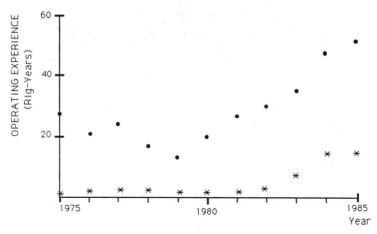

Fig. 1. Operating experience of mobile rigs in the UK sector of the North Sea: ●, all mobile rigs; *, jack-ups.

leads to significant permanent damage, is much less as can be seen from Table 1, and similar to that for fixed platforms. That only one in every eight collision accidents involving jack-ups leads to serious damage can be seen to be indicative of their ability to respond elastically to impact due to their high flexibility.

The prime causes of serious collision accidents in the North Sea have been found to be misjudgement and equipment failure,[2] as shown in Table 2. Weather does not appear to contribute significantly to serious accidents, perhaps due to decreased vessel activity during rough conditions.

By far the largest contributor to collision accidents are supply vessels, as shown in Table 3, and only about 2% of accidents are caused by passing vessels. There is also evidence[2] to suggest that the maximum

TABLE 1

Collision Accident Rates — UK Sector of the North Sea 1975–1985

Type of installation	Average collision accident rate (per rig-year)	
	All collisions	Serious collisions
Fixed steel jackets	0·188	0·033
Concrete platforms	0·108	0·031
Semi-submersibles	0·390	0·156
Jack-ups	0·279	0·033
All installation types	0·241	0·065

TABLE 2
Prime Causes of Collision Accidents which Led to Moderate
to Severe Damage — UK Sector of the North Sea 1975–1985

Prime accident cause	Percentage of reported accidents
Misjudgement	50
Equipment failure	29
Weather	4
Mooring	6
Other	2
Not specified	9

TABLE 3
Effect of Vessel Type on Collision Accidents
— UK Sector of the North Sea 1975–1985

Type of vessel	Percentage of accidents
Supply	70
Standby	9
DSV	11
Passing vessel	2
Other	8

collision velocities of supply vessels involved in the accidents presented
in Table 3 were of the order of 2 m/s, which is in agreement with DnV
design guidance.[3] In addition, as shown in Fig. 2, 95% of all collision
accidents in the southern North Sea involved vessels with displacements
less than 3500 tonnes. It would, therefore, appear appropriate that in
analysis simulating the response of jack-ups to collision impact, the
vessel displacement should be taken as 3500 tonnes, with a maximum
impact velocity of 2 m/s.

3 DEVELOPMENTS AND DESIGN GUIDANCE

Theoretical work on the behaviour of truss-legged jack-ups under impact
loading has been performed by DnV, using leg dimensions representative
of North Sea jack-up rigs,[4, 5] which led to the development of simple
design principles:

● The buckling strength of bracings should be sufficient to support the

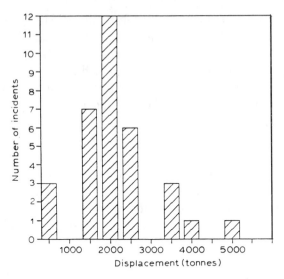

Fig. 2. Displacement of vessels involved in collision accidents in the southern North Sea.

chord such that this may develop a plastic mechanism of the shortest possible extent, i.e. one brace may buckle, but the next two bracings should remain intact.

• The ultimate bending strength of the chords should be sufficient to resist forces from bracings which have developed membrane tension yielding mechanisms

• Joints should have sufficient punching shear strength to resist compression loads developed due to the buckling of bracings and tension corresponding to yield stress in bracings.

The DnV analysis was performed using non-linear finite element modelling which suggested that 8 MJ of energy could be taken by a typical brace before it failed. It was also reported that an impact energy of about 5 MJ is required to generate a critical deflection in a leg chord, when the impact is against a joint, and with the direction normal to one of the sides of the equilateral triangle formed by the horizontal bracings joining the three chords.

Guidance on the design of jack-ups to resist accidental damage is contained in the DnV Rules for classification of Mobile Offshore Units.[3] Det norske Veritas also provide simplified procedures for estimating impact loads and vessel stiffness during accidental collisions,[7] which can be used in analysis.

Additional guidance is given in Classification Note 31·5,[8] concerning accidental ship collisions. It states that ship impact is most likely to cause local damage to one of the legs only, but the possibility of progressive collapse and overturning should also be considered. Since the overall flexural stiffness of jack-ups is considerably smaller than that of fixed jacket structures, elastic energy and response have a more significant role in resisting collision impacts. As a result Ref. 8 recommends that in most cases elastic analysis of jack-up impact response should produce adequate results.

An additional implication of this is that the plastic energy to be absorbed by the jack-up should be comparatively less than that which is proposed for fixed platforms.[1] However, there is very little information concerning the capability of a jack-up to withstand boat collision. Thus, it is not possible to state with confidence the level of accidental plastic energy which would be absorbed by a jack-up. The remainder of this paper attempts to provide such information.

4 ANALYSIS AND RESULTS

A three truss-legged jack-up, in 76·5 m of water, similar to that considered in Refs 4 and 5, formed the basis of the studies reported in this paper; it is illustrated in Fig. 3. The collision scenarios analysed were assumed to involve a 3500 tonnes displacement vessel, with impact force-indentation characteristics similar to those shown in Fig. 3 of Ref. 7. In the elastic analysis the ship was modelled as a concentrated mass and a linear spring element with an average stiffness of 10 MN/m. The analysis was performed as follows:

- A non-linear, elastic, dynamic analysis was carried out first using the finite element (FE) analysis package ABAQUS. This included added mass, taken to be 0·1 of the vessel displacement and damping effects, using a damping ratio of 0·06; the connections between the deck and the legs were modelled using appropriate spring elements, while the spud-can supports were modelled using rotational springs. A static analysis using this model produced results in very good agreement with those in Refs 4 and 5.
- The maximum impact force predicted by the FE analysis was then assumed to act quasi-statically. The method proposed in Ref. 9, extended to incorporate tensile membrane effects, was then used to estimate local denting and overall bending plastic deformations of the impacted member, as well as the corresponding, locally absorbed, plastic energy.

Three collision scenarios were considered:

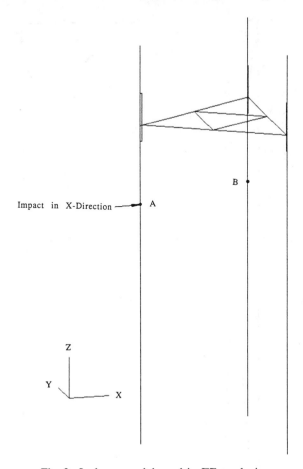

Fig. 3. Jack-up model used in FE analysis.

(a) Operational impact: the impact velocity was assumed to be 0·5 m/s. The FE analysis results are summarised in Fig. 4. As shown in Figure 4(a), the total kinetic energy involved in the collision is 0·48 MJ, while the maximum impact force is 2·6 MN. The effects of this on the jack-up chord, taken to have an outside diameter, D, of 1·0 m and thickness of 52 mm, are summarised in Fig. 5. It is apparent from scenario (a) Fig. 5(a) and from Fig. 7 that the chord develops localised dent damage only, with a maximum dent depth of 0·3% of the diameter, while the locally absorbed plastic energy, shown in Fig. 5(b), is only 0·005 MJ, or 1% of the kinetic impact energy. It is, therefore, reasonable to assume that the response of the jack-up to such an operational collision is elastic, with the vessel striking the chord and rebounding with most of its kinetic

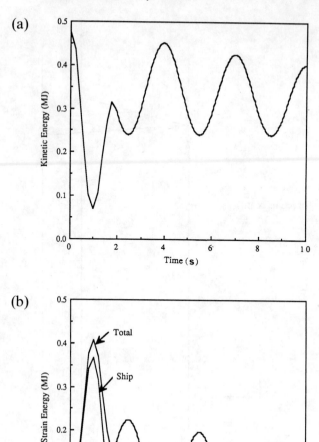

Fig. 4. Dynamic response of jack-up subject to impact by a 3500 tonnes displacement
vessel at 0·5 m/s: (a) kinetic energy; (b) strain energy.

energy retained, while the jack-up vibrates elastically, as shown in
Fig. 4(d).

(b) Accidental impact: the impact velocity was assumed to be 2 m/s, and
the FE analysis results are summarised in Fig. 6. In this case the kinetic
energy during the impact is 7·7 MJ, as shown in Fig. 6(a), and the

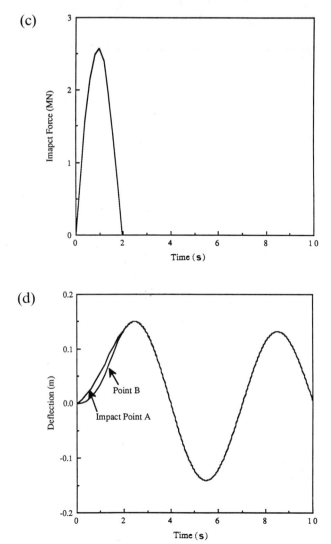

Fig. 4.—*contd.* (c) impact force; (d) impact deflection.

maximum impact force is 10·3 MN, as shown in Fig. 6(c). The localised chord response to impact is illustrated for scenario (b) in Fig. 5. The chord, again, develops only denting deformations, with a maximum dent depth of about 4·6% of the diameter, as shown in Fig. 5(a). The corresponding locally absorbed energy, shown in Fig. 5(b), is 0·315 MJ, which is only 4% of the kinetic impact energy. The jack-up response would, as a result, be expected again to be elastic. This is much in

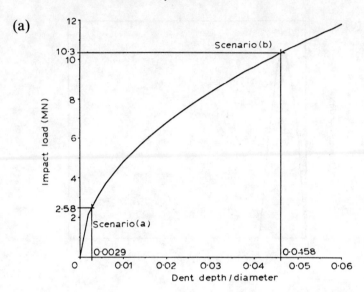

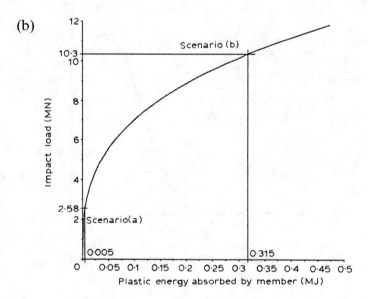

Fig. 5. Response of chord member to 3500 tonnes vessel impact (scenarios (a) and (b)):
(a) local denting damage; (b) absorbed plastic energy.

agreement with the recommendations of Ref. 8, and justifies the use of elastic analysis to simulate jack-up response to accidental collisions.

(c) Passing vessel collision: the impact velocity was assumed to be 6·0 m/s (12 knots), with a kinetic energy of about 70 MJ. The maximum impact

force developed as a result of such a collision is about 31 MN. The effects of this on a chord are to cause severe denting and bending deformations, of the order of 2·6% of the chord length, with a total absorbed energy of about 12 MJ.

The results from the analysis of these three collision scenarios are summarised in Table 4. It is clear from this table that for accidental collisions, as described in Ref. 1, the level of locally absorbed plastic

(a)

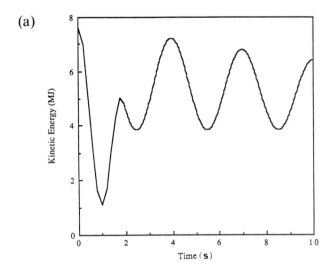

(b)

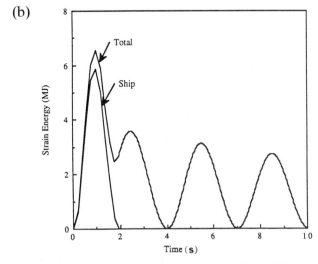

Fig. 6. Dynamic response of jack-up subject to impact by a 3500 tonnes displacement vessel at 2·0 m/s: (a) kinetic energy; (b) strain energy.

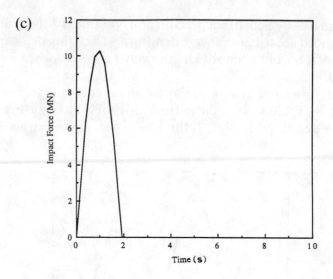

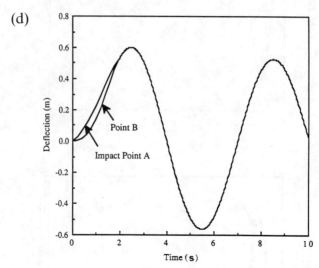

Fig. 6.—*contd.* (c) impact force; (d) impact deflection.

TABLE 4
Summary of Results from Collision Impact Analysis

Scenario	Impact velocity (m/s)	Kinetic energy (MJ)	Impact force (MN)	Local denting (d_d/D)	Overall bending (d_o/L)	Locally absorbed plastic energy (MJ)	EL/KE
(a)	0·5	0·48	2·6	0·003	0	0·005	0·02
(b)	2·0	7·7	10·3	0·046	0	0·315	0·04
(c)	6·0	70	31	0·41	0·027	12	0·17

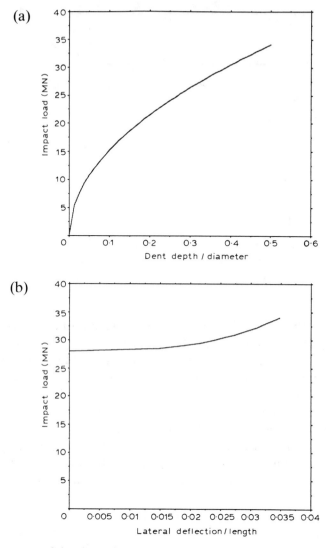

Fig. 7. Response of chord member to 3500 tonnes vessel impact (scenario (c) and ultimate resistance): (a) local denting damage; (b) overall bending damage.

energy by the jack-up is only a small fraction of the available kinetic energy. The resulting local chord deformations are also relatively small and would not cause any significant reductions in strength for the particular cases examined in this paper. However, fatigue considerations[6] may require early repairs, particularly for scenario (b), due to the very high local residual stress concentrations which accompany denting deformations.

(c)

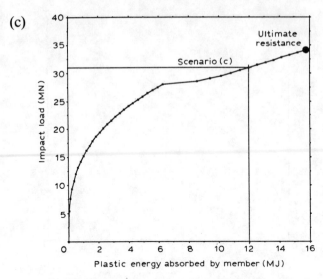

Fig. 7.—*contd.* (c) absorbed plastic energy

The results of further analysis to determine the maximum resistance of chord elements to lateral impact are summarised in Fig. 7. It can be seen from this that an ultimate impact force as high as 34 MN can be resisted by the chord which, however, develops extensive plastic deformations, absorbing in the process as much as 16 MJ of energy. However, its axial load carrying capacity would be reduced to a fraction of its original value, and the continued integrity of the leg and the jack-up would depend greatly on its ability to redistribute loading. Similar analysis of a bracing member, with an outside diameter of 0·4 m and thickness of 22 mm, indicates that it will be severed when the impact force exceeds 1·3 MN.

What is essential in such cases of extreme impact, due for example to passing vessel collision, is that it survives the impact, allowing safe evacuation and eventually repairs. This aspect of jack-up response to impacts, which North Sea experience shows can occur,[10] needs further investigation before more definitive guidance can be given.

5 CONCLUSIONS

What is very apparent from the analysis and results presented in this paper is the ability of jack-ups, due to their overall flexibility, to resist collision impacts much more resiliently than fixed platforms.

It would appear that their response to accidental impact is mostly elastic, developing only very small local denting deformations. However, the results from this analysis need to be extended further, by considering other classes of jack-ups and water depths, before definitive guidance can be produced on their assessment and design against accidental collision impacts.

REFERENCES

1. Department of Energy, *Offshore Installations: Guidance on Design, Construction and Certification*, draft fourth edition, October 1988, pp. 15.6–15.8.
2. Department of Energy, *Study on Offshore Installations: Protection against Impact*, background report, HMSO, 1988.
3. Det norske Veritas, *Rules for Classification of Mobile Offshore Units,* Hovik, Norway, 1984.
4. Pettersen, E. & Valsgard, S., Collision resistance of marine structures. In *Structural Crashworthiness*, Chapter 12, Butterworths, London, 1983.
5. Pettersen, E. & Johnsen, K. R., New non-linear methods for estimation of collision resistance of mobile offshore units. In *Proceedings of the 13th Annual Offshore Technology Conference*, Vol. 4, Paper 4135, OTC, Houston, Texas, 1981, pp. 173–86.
6. Hambly, E. C., Fatigue vulnerability of jack-up platforms. In *Proceedings of the Institute of Civil Engineers*, Part 1, **77** (1985) 161–78.
7. Det norske Veritas, Fixed offshore, installations — impact loads from boats, Technical Note, TN A 202, Hovik, Norway, 1984.
8. Det norske Veritas, Strength analysis of main structures of self elevating units, Classification Note 31·5, Hovik, Norway, 1983.
9. Ellinas, C. P. & Walker, A. C., Damage on offshore tubular bracing members. In *Proceedings of the IABSE Colloquium on Ship Collision with Bridges and Offshore Structures*, Copenhagen, 1983.
10. Offshore Engineer, *Brace to Give Damaged Rig a Leg up*, July 1988, p. 12.

Risk Analysis of Jackup Rigs

B. P. M. Sharples

Noble, Denton & Associates, Inc., 580 Westlake Park Blvd., Suite 750, Houston, Texas
77079, USA

W. T. Bennett, Jr

Friede & Goldman Ltd, 935 Gravier Street, Suite 2100, New Orleans, Louisiana 70112,
USA

&

J. C. Trickey

Noble Denton Consultancy Services, Noble House, 131 Aldersgate Street, London
EC1A 4EB, UK

ABSTRACT

*Increasing attention has been focussed in the North Sea and elsewhere on the
quantification of the risks of working in a hazardous environment: the
offshore world. The perception of risk with respect to mobile rigs has often been
vague and uninformed. This paper attempts to put the risks with respect to
jackup rigs into perspective by quantifying them and comparing them to other
risks.*

*This paper contains a few risk comparisons with fixed platforms, semi-
submersibles, and drillships. Historical casualties are used in an example to
show how a change intended to make an operation safer, may result in the
opposite effect.*

*Examining risks from losses due to environmental overload, the conclusion
is reached that jackups are very safe structures: there appears to be no jackup,
in the timeframe examined, that has been lost because of a deficiency in the
calculation methods currently in use by knowledgeable experts.*

Key words: risk, jackup, reliability, offshore platforms, safety, accidents,
failure probability.

INTRODUCTION

Risk Analysis is an important tool to decision making. Those who make decisions without the benefit of risk analyses may make decisions which increase the risk of what they are trying to avoid. This summer (1989) the US House of Representatives declared a moratorium on offshore drilling in environmentally sensitive offshore areas. The Interior Department has been banned for 1990 from making any lease sales or even preparing lease tracts in the continental shelf offshore in eight states, largely because of the Exxon Valdez oilspill. Referring to Fig. 1, it is obvious that tanker accidents represent a very much higher risk than oilspills from blowouts during exploratory offshore operations. Clearly, restricting offshore drilling will increase imports to the US by tankers which have approximately a 100 times higher risk. Such is the importance of risk analysis techniques as a tool to decision making.

Because of the attractiveness of jackups for both drilling and as early production systems, it is important to understand the risks clearly. Indeed as the statistics show, jackups may be 'safer' in many respects than the fixed platforms to which they are occasionally compared. Part of the reason is that they have advantages over fixed platforms in their ability to move off location. Underwater inspections can take place during moves; marine growth can be removed during moves; and maintenance and repair can take place in a shipyard setting, if required.

Noble, Denton & Associates has been extensively involved in site specific jackup risk assessments for owners and insurers for a number of years and has kept a detailed database on offshore incidents, particularly on jackup rigs. Friede and Goldman has been involved as designers in every aspect of offshore rigs and as such has acquired detailed knowledge of rig accidents and the safety features that need to be incorporated in the design of offshore rigs.

The jackup data in this paper reflects only the conventional drilling jackups. It does not address 'liftboats'. There are perhaps 150 liftboats regularly used on workovers, primarily in the Gulf of Mexico. In 1985 alone there were 29 incidents involving liftboats. Normally these unregulated vessels are designed to wave heights of only 3·5 m maximum and work in up to 23 m waterdepths, many miles from shore. There are currently approximately 75 with leg lengths of 27 m or more. They are operated by 2–4 man crews, are self-propelled and seek shelter whenever even a mild storm is forecast. They are not required to meet any safety standards and are not operated by the same community as jackup rigs, and therefore are not appropriate for inclusion in this study.

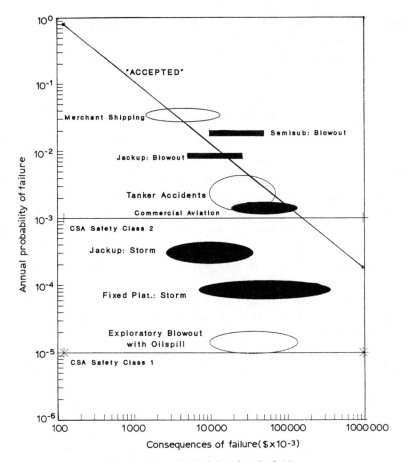

Fig. 1. Historical risk (after Ref. 18).

Often the methodology is to put forward a theory about some particular aspect of risk, assemble data and see if the answer makes sense. If it does not make logical sense, one has to examine more closely the question and the data that relate to it. If one had postulated the probability of dying on an offshore semisubmersible offshore Canada before the Ocean Ranger disaster, for example, the risk would have been small (if one based it on historical risks offshore Canada). Afterwards it was not so small. Reasoned argument would have told a risk assessor that the database was too small, and to relate the risk to activities elsewhere in the world for a more appropriate answer.

When looking at the activities that cause oilspills in an earlier study,[1] it became obvious that exploration blowouts producing oilspills were remarkably low. Blowout occurrences were quite high; but blowouts normally produce gas, not oil. The analysis as carried out in this paper does not purport to define all the answers, it merely tries to understand the accidents on jackups and also to shed some light on the areas of uncertainty with the use of risk analysis techniques.

ASSEMBLING THE DATA

Since every accident is in some way unique, classifying them into a reasonable number of types should be approached with caution, being careful in appropriately characterizing the incidents. Risk analysts typically do not discuss the difficulties in classification, how each incident fits into their selected categories. Concern over the subjective nature of the classification process and the fear that uncertainty in the process may lead others to criticize the conclusions is part of the reason. Those who use statistics without such detailed knowledge may easily be led to incorrect conclusions.

'Total loss' is a favorite category among some analysts. From a cursory glance it may be thought of as similar to 'death' in personnel statistics. The term 'total loss' can be very misleading. For example, a jackup may have a blowout which would cause damage to the extent of one-third of its value to repair; the controller of the well might decide to put explosives on the jackup and crater it, creating a total loss in order to more easily control the well. An alternative well controller might have taken a solution which did not involve creating the 'total loss' and then this event would not fit in the 'total loss' statistics. For another example, a jackup working over a fixed platform may suffer a sideways movement in a storm and by so doing find itself in the awkward position of not being able to jack down prevented by the presence of the fixed platform. Should it be characterized as a total loss due to soil failure?

The task is yet more arduous if one wishes to compare a loss of a fixed platform to that of a jackup. A fixed platform often represents a substantially greater investment than a jackup. A severe damage on a fixed platform will more likely be repaired than on a jackup since the jackup can be more easily removed from the location and replaced. Because of the need to recover the oil reserves with minimum downtime, fixed platforms are more often repaired even though replacement might be cheaper. The settling of the Ekofisk tank is not recorded in most statistics because there was no 'incident' yet it was a very costly experience, estimated at over $300 million.[2]

The key here is that in manipulating small amounts of data, as is the case with offshore unit data, the assessor must be extremely cautious in order not to create misleading conclusions.

For convenience jackups have been classified into 10 categories as follows:

Blowouts or Fires: Included here are spectacular blowouts such as the Mexico 2[3] and Ocean King,[4] those such as the Pool 55 (1987) that had a breakthrough while drilling and the Nan Hai 1 (1986) which resulted in no rig damage. Under Fires are categorized those fires which were not as a result of 'well problems' and include those with costly consequences such as the Western Triton 1[5] where an engine room fire caused a $5·5 million problem as well as those where fire caused little damage.

Soils: In this category both punchthrough and seabed problems were lumped together in Fig. 2, although they are separated in Table 1. Included here are incidents such as those where a jackup had to move off location due to scour, and the Houtech 1 (1982) where the mat was bent up while drilling over a boulder. One incident was recorded where volcanic action caused the loss.

The punchthrough incidents are certainly not comprehensive. Perhaps as low as 30% of those that have occurred have resulted in a 'recorded incident'. In some cases, no damage was recorded and in others, severe damage. There have been several cases when the jackup became a total loss because of a punchthrough.

Towing/Transportation: Incidents include both fair weather and rough weather towing. Damage ranged, from minor where the result of the incident was only that of being adrift for 24 h without a towline, to

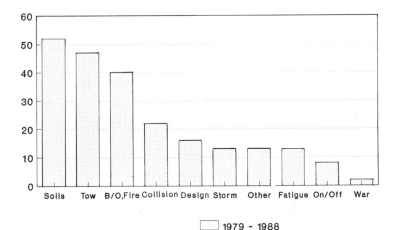

Fig. 2. Causes of jackup rig mishaps (total: 226).

TABLE 1
Cause of Jackup Rig Mishaps (1979–1988)

	1979–1983	*1984–1988*	*Total*
Towing	28	19	47
Punchthrough	16	27	43
Blowout	16	12	28
Collision	10	12	22
Design	4	12	16
Fire	8	4	12
Fatigue	4	9	13
Storm (jacked up)	9	4	13
Seabed	8	1	9
Other	3	5	8
Moving on/off location	4	4	8
Marine procedures	1	4	5
War	0	2	2
			Total = 226

accidents such as the Bohai 2 (1979) where the jackup capsized and claimed 72 lives. Included here also are such incidents as leg and bottom plating damage that occurred during dry transportation when the jackup was placed incorrectly on the cribbling and where the jackup was not appropriately secured for tow (e.g. Western Apollo 2 (1981) and Salenergy 5 (1981)), rigs where damage occurred offloading from the dry tow vessel (e.g. Sagar Vikas (1981)) and where the substructure shifted and was lost under tow (e.g. Vanguard 1 (1983)).

Collision: Incidents such as those where hull damage occurred from a tug hooking up, a supply boat banging into a leg on location, and those such as the Glomar Labrador 1 (1988) which was hit by the MV Irving Forest, are chronicled. There have been several units in collisions whilst in transit on dry transport vessels.

Other: In the category of 'other' incidents such as the Gulfdrill 4 (1984) where a section of the mat broke from unknown causes, the Diamond M Magellanes (1981) where a vacuum in the engine room caused several injuries, and those where damage arose from unknown causes, have been listed.

In Fig. 2 marine procedural errors are also shown in this 'other' category, though they are separated in Table 1. Mooring breakages when rigs were in port are classified under this heading (e.g. J. Storm 6

(1985), Zapata Scotian (1982)) and ballasting problems while not under tow (e.g. Montreal 2 (1987)).

Moving On/Off Location: Incidents categorized here are those where structural damage occurred when jacking. In some cases it was due to the wavestate exceeding that recommended for this operation.

Design: This category is reserved for those incidents where a repeated problem has occurred with a particular design, where spudcan vents were omitted by the shipyard, where lack of internal stiffening in the legs resulted in a distortion, and where equipment failed to perform its intended function in a way that might be expected in the industry. For instance, the Bill Bailey (1982/1983) incidents are categorized here. The incidents categorized here may not be truly 'designer' problems but may result from a different usage than the designer intended.

Fatigue: Chronicled under fatigue are such incidents as Ranger 1 (1979) where the mat/column connection failed due to fatigue and the Pool 145 (1982) where fatigue in dry tow was the cause. Some 'cracks in spudcans' have also been allocated to this category where they were thought not to be a 'design' problem.

Storm: A total of 13 jackups come into the category of storm damage and these are dealt with in more detail in the text. They consist of those incidents where a severe storm resulted in a problem while the unit was jacked up on location.

War: There have been two jackup rigs damaged by acts of war, in the period examined: the Maersk Victory (1986) and the Scan Bay (1988).

Arguably, the Ranger 1 categorized under 'fatigue' could alternatively be in the 'design' category. The evidence in each event has to be carefully weighed so as to not misconstrue the results. The selection of category often comes down to the subjective perception of the analyst.

EXPOSURE

In the estimation of probabilities the researcher must chronicle mishaps or anticipate the magnitude of the mishaps, which are then divided by the exposure. The exposure is the sum of the years of service for all jackup units for the period considered (the number of rigs × the number of years). Since the paper addresses risks to jackups, and comments on fixed platforms, semi-submersibles, and drillships, the worldwide exposure estimates for all four of these structure types were developed for

TABLE 2
Exposure of Rig Types (1979–1988)

Rig type	Rig years
Jackups	3 500
Semisubmersibles	1 150
Drillships	500
Fixed platforms	46 500

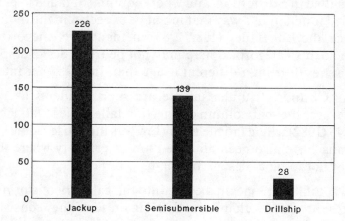

Fig. 3. Type of mobile rig with an 'incident', 1979–1988.

the 10 year period 1979–1988 and are reported in Table 2. The type of mobile rig with an 'incident' is shown in Fig. 3. The problem of selecting appropriate exposure levels is also not as straightforward as it might at first seem.

Referring to mobile rigs for example: should one include only those which are actively working offshore in calculating the exposure? Those that are stacked inshore might be said to have less risk. Underwriters customarily grant a lay-up premium reduction so they perceive the risk is reduced. Certainly the exposure to blowouts is less if the unit is not drilling. Those that viewed the 35 jackups stacked in Sabine Pass with large ships steaming by might not have felt they had less exposure from the collision risk. Likewise they would have had to survive a hurricane risk, although in the shallow water the waveheights would have been significantly below the design level for all of them. Certainly some of the rig damages chronicled are those which occurred more often when in a shipyard (fire being a high risk in the shipyard; breaking moorings is also a risk to a vessel with a high wind area and little in the way of equipment to equalize tensions in the lines required to hold it inshore).

When considering exposure levels appropriate for fixed platforms, equally, one has hurdles to cross. What should one include as fixed platforms? What about the inland wellhead structures or those platforms in 'state' waters. One might consider omitting the offshore units which are abandoned or produce below certain levels. The number of fixed structures in the world has to be estimated in any case since there are few bodies that accurately chronicle them.

One could argue that the exposure level calculated for jackups is not reporting similar 'risk exposure' as that calculated for fixed platforms. After all, jackups may often be used well below their design capability on shallow water locations, or in relatively benign environments. Fixed platforms almost exclusively are designed for the extreme storm at the location at which they are installed.

To a large extent, the offshore oil world's history in fixed platforms has been to a great extent in the Gulf of Mexico, where there are many fixed structures. If several platforms are lost in one hurricane, is that one 'incident' or several?

In reporting the exposure levels in Table 2, judgement has been used. The numbers can be argued up and down. Many steps along the way in risk analysis involve investigator judgements, and this is just one of them.

FOUNDATIONS

Jackup risk related to foundation failure merits detailed examination. Many foundation failures are the result of procedural errors and a rational decision to take an increased risk. The most common jackup foundation failures are those associated with:

Punchthrough of the footings	70%
Foundation failure due to severe storm loading conditions	16%
Scour around the footings	5%
Interaction with previous footprints	3%
Other	6%

The apparent incident of punchthroughs is an increasing problem. The term 'apparent' is used because historically only incidents have been reported where losses occurred. Interest has only been focussed on this problem since changes in jackup design have made some jackups more vulnerable to this damage. Of the 43 incidents recorded in the period about 50% had some serious damage. Those not damaged are primarily those whose legs have a high shear and bending moment capability compared to the weight of the elevated hull. It appears some jackups currently in use have a higher hull-weight-to-leg-strength ratio than some of the older jackups. This results in a lesser ability to withstand the rapid penetration events that may occur from time to time.

Further examination of the data indicates there may be a trade-off in risk. Design of legs to minimize wave forces often results in concepts which increase risk of damage in a situation of rapid penetration during preloading. Damage can be avoided in jackups with this characteristic merely by preloading at minimum airgap in a situation where rapid penetration is a possibility. Airgap is often defined differently by the designer and those on the jackup. One definition, often a designer's, may present the distance between still water level at the bottom of the tide and the underside of the hull as the 'airgap'. The operations person may believe the 1·5 m recommended airgap for preloading may mean the distance between the bottom of the hull and the wave tops at the highest tide during the preloading procedure. These differences in definition alone can result in a 1·5–2·5 m difference in airgap in a no-tide situation.

Ideally, if there is any uncertainty about the foundation at a location, corings should be available beforehand to indicate the likelihood of punchthrough and hazard mitigation precautions developed as appropriate. If penetration on location is significantly different from that predicted, clearly additional coring should take place before the jackup is taken to full airgap. 'Improper procedures' have often been the cause of what becomes classified as 'foundation failure'. Occasionally, 'stipulated procedures' cannot be followed. The Rio Colorado 1 (1981) was involved in a punchthrough offshore Argentina. The operating procedures for that jackup specified a 1·8 m airgap. Due to lifting capacity limitations, typical of jackups, it was not permitted to jack with the additional weight of the preload, and was therefore not able to follow the 10·7 m tidal range at that location. The type of design was beneficial for this particular situation and fortunately only very minor damage was sustained.

Mudslides are a foundation problem in certain areas of the world. The 'Harvey H. Ward' (1980) was lost in Hurricane Allen due to a mudslide.

There is a clear increased risk attached to operating in Delta areas. Those taking the decision to explore in those areas cannot avoid the risk: they can merely minimize it by selecting the most suitable location on an area least likely to be an active lobe. Use of a semisubmersible in such circumstances also does not circumvent the problem. The Western Pacesetter 3 was involved in a mudslide in November 1980 and the well was lost.

Mat jackups, by design, often work in areas of soft soil. When they do so there is a risk of sliding failure in the event of a storm. There have been many of these mishaps. In Hurricane Allen (1980) alone, for example, Salenergy 1, Teledyne 17, Sabine 1, Fjelldrill, J. Storm 7, Mr. Gus 2, and Pool 50, all slid sideways. Calculations can easily show that the factors of safety on sliding in a minimum 10-year hurricane are often less than 0·8–0·9. If oil companies wish to drill in soft mud areas, it is a risk that must be accepted.

The problem jackup designers face is that of maximizing the utility of the jackups on all possible foundations whilst minimizing the risk of failure in any one type of foundation. With the large variety of locations under which oil is found or anticipated, this is a very difficult problem. There have been several jackups where high load bearing footings have been designed to maximize penetration, and minimize any possibility of sideways leg slip leading to a change in load sharing in the legs, which later led to major modifications in order to compete commercially.

The statistics do not reflect near misses. One case of a near miss occurred on a jackup with particularly large diameter flat-bottomed footings where the rig jacked up on a hard sand bottom with a localized slope such that at least one leg's vertical reaction was taken toward the edge of the spudcan. This was not noticed during preloading. The preloading occurred at an usually high airgap because of extreme tides. The preloading time was extended because of a safety precaution of only preloading through the top half of the tide: the preload was 'held' through the bottom half. Unfortunately the weather increased during preloading such that jacking down was not a viable option. The high leg reaction forces coming through the edge of the can put the legs in a geometric bind. The jacking system was not powerful enough to jack through the bind. Later, fortunately, the force/motion from the increasing waves/scour or some alternate mechanism caused the load to distribute and the rig jacked up just as the waves were reaching the underside of the hull.

Table 3 presents methods of minimizing risk for various typical foundation problems.

JACKUP RISK

In calculating the probability of 'jackups damaged whilst jacked-up during storms' one could take the 13 listed in Table 4 and divide by the 3500 rig years in Table 2 for a result of 3.7×10^{-3}. Alternatively, it is likely that some of the time a jackup may operate well below its capability and thus is not operating at as much risk; for some that could be, for example,

TABLE 3
Risk Reduction in Foundation Failures

Risk	Method of minimizing risk
Punchthrough	— Shallow seismic survey — Coring data — Preload below minimum airgap — If penetrations do not correspond to predicted, core from rig
Scour	— Know surface sediments and currents — Inspect footing foundations regularly — Gravel bag/grass mat the footings where anticipated
Mudslides	— Obtain qualified expert's report
Gas pockets	— Shallow seismic
Faults	— Shallow seismic
Metal or other object — sunken wreck anchor, pipelines	— Magnetometer and side scan sonar — Diver's walk
Local holes (depressions) in seabed, reefs, pinnacle rocks or wooden wreck	— Side scan sonar — Diver's walk
Sliding failure	— Surface coring — Pile the footing(s)
Bearing failure	— Coring data
Legs stuck in mud	— Coring data — Consider change in footings/overcan jetting, etc.
Excessive storm penetration	— Ensure adequate soil strength and jackup preload capability
Footprints of other jackups	— Evaluate location records, consider filling if necessary

50% of the time. If the intended use of the statistics is to quantify the risk to a jackup which is being operated at its design limits it would be proper, under these assumptions, to only use the comparable exposure level leading instead to a risk of $7\cdot4 \times 10^{-3}$. Since the 50% used in the example has not been adequately derived, it is not used further in this paper, and just the gross statistics are utilized in further discussions.

If one examines the jackups listed in Table 4 and omits those where an increased risk of sliding was accepted (mat units), and those where very cursory site specific assessments would have shown a failure would be

TABLE 4
Jackup Rig: Storm Damage

Year	Name		Damage
1980	Harvey H. Ward	Mat	Mudslide in Hurricane Allen
1980	Salenergy 1	Mat	Shifted position in Hurricane Allen
1980	J. Storm 7	Mat	Shifted position due scour in Hurricane Allen
1980	Teledyne 17	Mat	Shifted position due scour in Hurricane Allen
1980	Sabine 1	Mat	Damaged substructure, wellhead and BOP in Hurricane Allen
1980	Fjelldrill	Mat	Tilted during Hurricane Allen — damaged
1980	Western Triton 4	3 Leg	Bent drive stacks and BOP in Hurricane Allen
1980	Dixilyn Field 81	3 Leg	Took on additional penetration and collapsed in Hurricane Allen
1983	Mr Gus 2	Mat	Slid 2·5 m off location in 8 m seas. No damage.
1984	J. Storm 17	Mat	Hit by waterspout. Helicopter lost — cantilever moved.
1985	Pool 50	Mat	Slid off location in Hurricane Danny — leaned toward a fixed structure — couldn't jack down.
1985	Dyvi Beta	3 Leg	Water tower broke off in storm
1985	Penrod 61	3 Leg	Took on additional penetration and collapsed in Hurricane Juan

Note: Hurricane categories range from 1 (minimal) to 5 (catastrophic)

Year	Hurricane	Category
1980	Allen	3
1985	Danny	1
1985	Juan	1

extremely likely, one can conclude that there have been no jackups that had a major failure due to the calculation methods. The exposure levels of jackups have been small compared to fixed platforms (3500 *vs* 46 500). Therefore one cannot show by statistics that the methods used[6] are 'safe'. That would take an exposure level of at least one to two more orders of magnitude. One is therefore left with only being able to say that jackups are as safe in a storm as the exposure level $1/3501 = 2.8 \times 10^{-4}$ when jacked up. It is therefore concluded that because of a limitation in the data, the risk can only be said to be less than 2.8×10^{-4}.

In order to be sure that by restricting the timeframe to the last 10 years, important data were not excluded, a check was made of the period 1969–1978. No failures due to storm overloading on jackups, that meet a minimum criteria in the elevated condition, were found during that period.

JACKUPS VERSUS FIXED PLATFORMS

Jackups are sometimes criticized as being designed to a lower standard than fixed platforms. Figure 4 chronicles rig mishaps for both types. In reviewing the mishaps case-by-case several interesting aspects come to the fore.

When a mishap occurs on a fixed platform, it is more often than not very much more serious. Fires and blowouts are by far the most risky hazards for fixed platforms and the consequences from these are normally large.

Towing, historically, is a hazardous operation for both mobile rigs and fixed platforms. Mobile jackups are towed several times a year: fixed platforms only once in their lifetime. The frequency of towing accidents for the mobile rigs is therefore expected to be higher.

Punchthrough is not a problem for fixed platforms: because of the expense of the investment to the oil companies, full soil investigations are usually carried out ahead of time for fixed platform installations and they are not preloaded to their maximum environmental ratings during installation. Thorough geotechnical information is often not developed for siting mobile rigs.

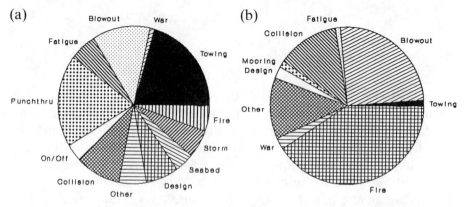

Fig. 4. Cause of rig mishaps — jackup *vs* fixed platform: (a) distribution for jackups; (b) distribution for fixed platforms.

There have been some spectacular collisions with fixed platforms recorded, such as a supply boat bringing a manned platform into port on its aft deck (State Command, 1976)[7] and one where a tanker impaled itself on a partly constructed fixed platform (Eugene Island 361-A).[8] The Glomar Labrador 1 (1988) hit by the MV Irving Forest is probably the most severe jackup collision damage in recent years. It is believed that the extent of some of the more spectacular collisions is merely a result of more exposure in the case of fixed platforms.

A comprehensive study by Gulf Research and Development[9] established the distribution of blowout frequency in the US Gulf of Mexico which is believed to be typical worldwide. Their study was based on 98 blowouts between 1955 and 1980 as follows:

Platform	40%
Jackup	13%
Semisubmersible	5%
Drillship	2%

Clearly, these figures must be used carefully since they have not been related to the exposure levels.

Looking in detail at the comparison of risk related to environmental extremes it is necessary to review, for jackups and fixed platforms, the storm casualties. Table 4 shows those for jackups and Table 5 for fixed platforms. It is interesting to note from the investigation of the last three platforms listed that the fixed platforms were underdesigned.[10] Due to subsidence and an increase in wave height parameters appropriate for design, the Ekofisk Platform was damaged.[2] It is interesting to reflect that once a fixed platform is installed, if waveheights appropriate for design in the geographical area change there is a major problem. For jackups, when such an occasion arises, one can merely jack up to a higher airgap

TABLE 5
Fixed Platform: Storm Damage

Year	Name	Damage
1983	Esher 4	6 m waves eroded artificial island ($46 million).
1984	Ninian N	Bracing severed at −43 ft (15 m) in 60 ft (20 m) seas.
1985	Platform A	Structural collapse. Sank in Hurricane Juan.
	S. Timbalier 86	Under-design and corrosion.
1985	S. Pelto 19	Structural collapse. Sank in Hurrican Juan.
	OBM Header	Under-design and corrosion.
1985	S. Pelto 19	Structural collapse. Sank in Hurrican Juan.
	Well Protector	Under-design.

next time it comes onto the location or use a harsher environment jackup if at or beyond its design condition. Alternatively, the jackup can be used directionally, if appropriate, or seasonally. No such options exist for fixed platforms; therefore they tend to be used even though they may not be up to the current industry standard. Frequently oil companies apply more refined and sophisticated calculation techniques to increase the capacity of the platform to resist increased environmental loadings or justify the continued use of a damaged or aging platform. These procedures reduce the safety factors and therefore increase the risk factors for those platforms. Currently several fixed platforms are being reviewed for suitability in the southern North Sea, Australia and Gulf of Mexico.[11, 12]

Referring to Table 4, typical mat jackups are designed to work in areas of weak soil. Working in these areas of soft soil brings with it an increased risk. As indicated above, typical safety factors are 0·8–0·9 for sliding in a 10-year return period hurricane. If one uses these jackups in that situation, one should expect higher failure rates.

Particular precautions can be justified for fixed platforms installed in these areas because of the increased capital that can be generated once a prospect is a reality.[13, 14]

The Penrod 61 and Dixilyn Field 81 are interesting casualties. These rigs reflect characteristics of some of the older jackups in the industry. Figure 5 shows, in conceptual form, typical capabilities of these jackups. Plotted here are vertical leg reactions for 'not untypical' designs to illustrate the problem. Designers typically design jackups to a certain wind, wave and current criteria for a specific waterdepth. Clearly it would be impossible to anticipate these parameters at every location in which a design might operate during its lifetime when it changes operating location several times per year and even operational areas from time to time. Most of the jackups used in the Gulf of Mexico have the capability to withstand a winter storm loading and generally the 10-year return period storm at typical operating waterdepth in the hurricane season. Because these jackups are able to be readily demanned, they are generally not capable of withstanding a 50-year storm return event in hurricane season.

Figure 5 shows a progression of design in illustrative form from some of the early jackups, Design A, to current practice, Design C. Leg reactions corresponding to initial and final leg penetrations before and after preloading are given. These can be compared to a winter storm or to a 10-year or 50-year return period hurricane.

When comparing risk levels in the Gulf of Mexico there is the additional complication that there are no generalized guidelines for waveheights. Consultation of the API Recommendations[15] leads one to

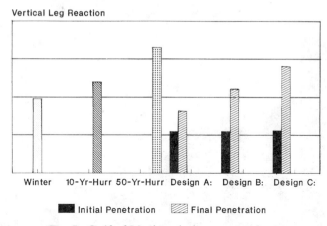

Vertical Leg Reaction

Winter 10-Yr-Hurr 50-Yr-Hurr Design A: Design B: Design C:

■ Initial Penetration ▨ Final Penetration

Fig. 5. Gulf of Mexico: design progressions.

the conclusion there is a diversity of opinion about waveheights. For example, in 90 m of water, opinions vary between 19·5 m and 24 m at apparently similar locations. Certainly, some platforms exist in the Gulf of Mexico designed to lower values since previous API Guidelines[16] show an opinion on wave heights as low as 17·5 m for the same 90 m water depth. Since there is no uniform guideline adhered to in the Gulf of Mexico, fixed platforms and jackups may both exist in a higher risk state than the norm if they are designed or used in areas only appropriate for the lower wave height.

Referring to Table 5 and Table 2, if the artificial island is omitted since islands are omitted in the exposure statistics, one could propose a failure rate due to storms for fixed platforms of $4/46\,500 = 8\cdot6 \times 10^{-4}$. Simply because jackups have had 1–1/2 orders of magnitude less experience than fixed platforms, for the time being it is not permissible to use statistics to show that jackups are safer.

JACKUPS VERSUS SEMISUBMERSIBLES

Causes of semisubmersible rig mishaps are given in Fig. 6. Mooring is the highest risk to semisubmersibles.

It is interesting to reflect that with the configurations put forward for fixed platforms in deep water depths (90 m) in the North Sea, semisubmersibles have been put forward in informal discussions as safer alternatives to jackup rigs. Jackup rig safety has been questioned as it relates to dynamics in deep water, largely because of a lack of interest/funding until recently, in developing quick inexpensive assessment techniques for deep water applications verified by experimental evidence in extreme storm situations.

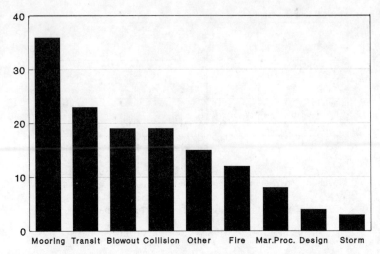

Fig. 6. Causes of semisubmersible rig mishaps, 1979–1988 (total: 139).

For a jackup on such a deepwater location one might rationalize a risk level of (in round figures) 5×10^{-4} if suitable site assessments confirm the acceptability of a location.

For a semisubmersible the historical risk of a mooring failure is in excess of $36/1150 = 3 \times 10^{-2}$. Additionally, moorings are not typically designed for the skewed system necessary to work alongside a fixed platform in the North Sea; and the mooring systems are often designed to only withstand a 10-year return period event. In working as a tender to a fixed platform one would most probably have a bridge connection between the platform and the semisubmersible. There have been at least two events in the last two years where the bridge has collapsed on semisubmersible accommodation units. In one case, the bridge, heresay has it, scraped the barnacles off the export pipeline. Thus in an effort to avoid damage due to possible dynamics problems in jackups one could argue that the solution of having the work carried out by a semisubmersible could result in about 2 orders of magnitude (100 times) more risk.

Other comparisons can be made between semisubmersibles and jackups. Blowouts are a hazard of the oil patch. In the period examined, 1979–1988, there were 19 blowouts on semisubmersibles and 28 on jackups in the database. Thus the risk is $19/1150 = 16 \cdot 5 \times 10^{-3}$ for semisubmersibles and $28/3500 = 8 \times 10^{-3}$ for jackups. The conclusion one can reach is that a semisubmersible is twice as likely to have a blowout as a jackup based on historical statistics.

The reason for this has not been examined in detail in this paper, although several possibilities come to mind. Fluid control methods being more difficult on semisubmersibles, higher pressure gas zones drilled by semisubmersibles, and differences in well control systems are all areas wherein an explanation may lie. The risk of an incident with a semisubmersible based on the total number of incidents divided by total exposure is $139/1150 = 12 \times 10^{-2}$.

JACKUPS VERSUS DRILLSHIPS

Causes of drillship mishaps are given in Fig. 7. Blowouts are the highest risk incident followed closely behind by collision. The probability of an 'incident' is $40/500 = 8 \times 10^{-2}$ which is between that of a semisubmersible and that of a jackup.

A cursory examination was made to the risk of a blowout associated with drillships to add to this comparison. There were seven blowouts chronicled leading to a frequency of $7/500 = 14 \times 10^{-3}$. This again leads to the conclusion that the risk of a blowout using a drillship is much more in line with semisubmersibles. Jackups again appear to come out as a significantly better risk.

RISKS IN PERSPECTIVE

Offshore risks are summarized in Table 6.

In order to give some guidance on acceptability of risk, the Target

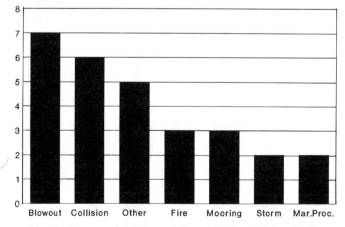

Fig. 7. Causes of drilling mishaps, 1979–1988 (total: 40).

Annual Probability Levels recommended by the Canadian Standards Association[17] are given below (Table 7).

These have also been presented in Table 8 together with a collection of other events that have been taken from various sources.

Figure 1 shows another method of visual presentation giving a perspective on risk in two dimensions. The Graph type was published by Whitman in 1984.[18] Several items are imposed on this graph to attempt to put the risks in perspective.

The recent apparent frequency of airline disasters caused a re-examination of the statistics for airline accidents quoted in the literature.[18] Airline travel has often been depicted as the safest form of transportation. On the basis of 'per mile travelled' this is probably a fair

TABLE 6
Offshore Risks

Risk of an incident	
Jackup	6×10^{-2}
Semisubmersible	12×10^{-2}
Drillship	8×10^{-2}
Risk of blowout	
Jackup	8×10^{-3}
Semisubmersible	$16 \cdot 5 \times 10^{-3}$
Drillship	14×10^{-3}
Risk of jackup major damage due to storm (all site precautions taken):	less than $2 \cdot 8 \times 10^{-4}$
Risk of fixed platform damage due to storm:	$8 \cdot 6 \times 10^{-5}$

TABLE 7
Canadian Standards Association: Safety Classes and Probability Levels

	Consequences of failure	*Target annual probability of failure*
Safety class 1	Great risk to life or high potential for environmental pollution or damage	10^{-5}
Safety class 2	Small risk to life and low potential for environmental damage	10^{-3}
Serviceability	Function (use or occupancy) impaired	10^{-1}

TABLE 8
Comparative Risk (in order: highest to lowest)

Canadian standards: serviceability	1.0×10^{-1}
Blowout: semisubmersible	16.5×10^{-3}
Offshore worldwide blowouts (1955–1980)	8.3×10^{-3}
Blowout: jackup	8.0×10^{-3}
Tanker accident:	2.3×10^{-3}
Air travel:	1.5×10^{-3}
Canadian standards: safety class 2	1.0×10^{-3}
Jackup major damage due to storm (all precautions taken):	less than 2.8×10^{-4}
Motor vehicle accidents:	2.4×10^{-4}
Police killed in line of duty:	2.2×10^{-4}
Fixed platform major damage due to storm	8.6×10^{-5}
Exploratory offshore blowout with oilspill:	1.4×10^{-5}
Canadian standards: safety class 1	1.0×10^{-5}

characterization. However, since 68% of airline accidents happen at takeoff (23%) and approach/landing (45%) a per trip risk could be a better characterization. The risk methodology used for jackups was also applied to commercial aviation. There are some 8000 commercial jets in the free world and approximately 2500 in the US. The disaster/accident rate is approximately 10–15 per year and 2–3 per year respectively. This results in a typical statistic of 1.5×10^{-3}. Thus, if commercial aviation is compared on the same basis as jackups (no. of accidents divided by no. of airplanes) the conclusion is reached that jackups in an elevated mode may be safer than commercial air travel. Certainly, that comparison is not out of line with what knowledgeable experts on jackups 'feel' about being on board them.

Each reader is left to draw his own detailed conclusions from the data presented by this method.

CONCLUSIONS

The risk of jackup rig operations have been chronicled in this paper. Several of those risks developed are reported in Table 6. These are compared to other risks in Table 8. The highest risks are from moving rigs from location to location and placing them onto their new foundations. The risk from storm damage if a site assessment is carried out has been demonstrated to be less than 2.8×10^{-4}. The site assessment methods required to produce such a low frequency of accident are comparatively very simple and very inexpensive.

The nuclear industry examines risk based upon cost of dollars to save lives or minimize damage.[19] If the jackup industry examined risks by the same methodology one could conclude, notwithstanding that simple site assessments should be carried out, that if dollars are spent, they should more effectively be used on problems associated with towing and moving onto location, and then on safety features that decrease the number of blowouts and collisions in the jacked-up mode.

In examining the risks from losses due to environmental overload, the conclusion is reached that jackups are very safe structures: there appears to be no jackup, in the timeframe examined, that has been lost because of a deficiency in calculation methods currently in use by knowledgeable experts.

REFERENCES

1. Sharples, B. P. M., Kalinowski, D. K., Tidmarsh, G. & Stiff, J. J., Statistical risk methodology: application for pollution risks for Canadian Georges Bank drilling program. *Offshore Technology Conference*, OTC 6082, May 1989.
2. Bleakley, W. B., Subsidence: what is it, what causes it, and what can it cost? *Petroleum Engineer,* November (1986) pp. 31–36.
3. Anon., Pemex removing wrecked jackup in Gulf. *Petroleum Engineer International,* June (1988).
4. USGS, Investigation of August 1980 blowout and fire, Lease OCS-G 4065 Matagorda Island Block 669, Gulf of Mexico off Texas Coast. USGS Open File Report 81–706, 1981.
5. Le Blanc, L., Tracing the causes of rig mishaps. *Offshore,* March (1981) 51–62.
6. Sharples, B. P. M., Gorman, M. T. & Firstenberg, C. E., Interpreting meteorological and marine surveying criteria for jackup rig design, selection and placement. *Second International Conference on Offshore Safety,* March 1986.
7. *United States Coast Guard,* No Comment. In *Proceedings of the Marine Safety Council,* USCG, **34**(1) (1987).
8. US Tankship S/S Texaco North Dakota and Artificial Island EI-361-A Collision and Fire Gulf of Mexico, August 1980. NTSB-MAR-81-4.
9. Warlick, W. P., Goodwin, R. J., Teymourian, P. & Krieger, W. F., Analysis of accidents in offshore operations where hydrocarbons were lost. Houston Technical Services Center, Gulf Research and Development, 1982.
10. Minerals Management Service, *Investigation of October 27–28, 1985 Structural Failures.* OCS Report MMS 87-0075, MMS, US Dept of the Interior, Gulf of Mexico, 1985.
11. Bea, R. G., Puskar, F. J. & Spencer, J. S., Development of AIM (Assessment, Inspection, Maintenance) programs for fixed and mobile platforms. *Offshore Technology Conference,* OTC 5703, Houston, 1988.
12. Cottrill, A., Props put new life into ageing Australian platforms. *Offshore Engineer,* April (1988) pp. 24–30.

13. Bea, R. G., Wright, S. W., Sircar, P. & Niedoroda, A. W., Wave-induced slides in South Pass Block 70, Mississippi Delta. *ASCE Annual Convention,* Hollywood-by-the-Sea, October 1980.
14. US Department of Interior, Environmental information on hurricanes, deep water technology and Mississippi Delta mudslides in the Gulf of Mexico. BLM Open File Report No. 80-02, 1982.
15. American Petroleum Institute, *Recommended practice for planning, designing and constructing fixed offshore platforms.* API Recommended Practice 2A (RP 2A). API, Washington DC. Sixteenth Edn, April 1986.
16. American Petroleum Institute, *Recommended practice for planning, designing and constructing fixed offshore platforms.* API Recommended Practice 2A (RP 2A). API, Washington DC. Eleventh Edn, Jan. 1980.
17. Canadian Standards Association, *Guidelines for the Development of Limit States Design,* CSA Special Publication S408, 1981.
18. Whitman, R. V., Evaluating calculated risk in geotechnical engineering. *J. Geotechnical Eng.,* **10** (1984) 143–88.
19. Siddall, E., A rational approach to public safety — an interim report. *Conference on Health Effects of Energy Production,* Chalk River, Canada, 1979.

Extreme-Response Analysis of Jack-Up Platforms

H. Kjeøy, N. G. Bøe & T. Hysing

Det norske Veritas Classification A/S, Veritasveien 1, N-1322 Hoevik, Norway

ABSTRACT

The extreme wave and current response of a typical North Sea jack-up drilling platform has been calculated using nonlinear, stochastic, time-domain analysis techniques. A statistical treatment of the response-time histories has been performed. Appropriate probability distributions have been fitted to simulated response amplitudes and extrapolated to determine extreme values in storms of 3 or 6 h duration. Nonlinear geometric effects and dynamic effects are also addressed.

The results of the nonlinear time-domain analyses are compared to the results of more simplified methods used in jack-up designs.

Key words: jack-up platforms, extreme response, irregular wave loads, time-domain analyses, dynamic behaviour, non-linear effects.

1 INTRODUCTION

The horizontal stiffness of a jack-up platform is typically an order of magnitude less than the stiffness of a corresponding jacket structure. The important consequences of low stiffness are as follows.

— dynamic effects become important, since the natural period of the platform increases with reduced stiffness. Typically, the fundamental mode of jack-up platforms corresponds to a natural period in the range of 4–8 s, which coincides with wave periods containing significant amounts of energy.

— Second-order geometric effects increase with reduced stiffness due to hull mass lateral displacements.
— The relative velocity between the structure and the water particles changes, and will directly affect the Morison type wave loading on the structure (see Section 3.2) as well as the hydrodynamic damping properties.

In spite of the above effects, jack-up platforms have traditionally been analysed according to the static methods normally used for jacket structures. In areas where the wave current loading contributes only a minor part of the total loading, such procedures may give acceptable results provided the parameters and assumptions used in the analysis are assessed in a reasonably conservative manner. As the use of jack-up platforms extends to more harsh environments, it cannot, however, be considered satisfactory to employ methods of analysis which do not account for effects which represent important characteristics of the structure being analysed.

As a consequence, in 1982 Det norske Veritas published a Classification Note,[1] which gives guidance on a deterministic (regular wave) analysis of jack-up platforms where dynamic and second-order effects are accounted for in an approximate and simplified manner. Although such a procedure cannot reflect the real behaviour of a jack-up platform in irregular (stochastic) seas, it is considered to be a useful tool, providing safe results without being unduly conservative for realistic combinations of structural configurations, water depths, and severity of hydrodynamic loading.

Stochastic, nonlinear, time-domain analysis methods are unique in the sense that the real behaviour of the structure can be simulated and it is possible to account directly for all of the important effects due to dynamic, stochastic, and nonlinear response. The methods, however, require very long computer times and the interpretation of the results may not be straightforward. Hence, these methods should not be considered as a 'standard' tool for design analyses. Instead, the methods may be used for the final verification of special designs and/or for calibration of, and comparison with, simpler methods of analysis.

The purpose of this paper is to compare the results of various methods for global analysis of jack-up platforms in the elevated condition. Extreme-wave conditions only are considered. The effect of wind, although important for design, has been disregarded for the purpose of this study.

The following methods have been considered for comparison:

— static, regular-wave analysis without correction for dynamic and second-order effects;

— static, regular-wave analysis with applied correction factors to account for dynamic and second-order effects;
— static, nonlinear, regular-wave analysis;
— dynamic, nonlinear, regular-wave analysis;
— stochastic, nonlinear, time-domain analysis.

The analyses have been carried out for a typical North Sea jack-up platform in 75 m water depth employing the SESAM/FENRIS computer systems.[2-4] The results are strictly valid only for the example platform subjected to the given conditions. However, the phenomena which are demonstrated are believed to be rather typical for these types of structures.

2 DESCRIPTION OF EXAMPLE CASE

2.1 Platform characteristics

The example platform consists of a triangular hull structure with three lattice-type legs, each with triangular cross-section. The main dimensions are

   — longitudinal leg centres   51 m
   — transverse leg centres   55 m
   — leg chord centres   10 m
   — leg bay height   5 m (K-braced)

The platform weights are

   — total elevated hull   (8800 tonnes)
   — one leg, excluding can   (850 tonnes)

The elevating system consists of a four-high opposed-rack system. The guide system is parallel to the rack plate with the guide vertical centreline distance of 15 m.

2.2 Structural model

The triangular lattice-type legs have been idealised by equivalent cylindrical legs with the stiffness corrected according to the procedure given in Ref. 1. A comparison with a detailed leg model including all leg elements has also been carried out showing good agreement between the two models. The leg boundary conditions at the sea floor are constrained against translations in the x, y and z directions but free to rotate about the same axes. Thus, no effort has been made to model the complex leg–soil interaction aspects. The leg boundary conditions at the hull have been simulated by linear springs of appropriate stiffnesses at the upper and lower guides and at the level of the elevating system.

The hull is represented by a suitable membrane element model accounting for the stiffnesses of the deck, sides, bottom, and main structural bulkheads.

A plot of the computer model which was generated by the SESAM system[2] is shown in Fig. 1. The total degrees of freedom for this model is 504.

2.3 Leg hydrodynamic properties

The hydrodynamic properties of the equivalent leg have been calculated based on coefficients traditionally used for regular wave analysis without any allowance for marine growth. Thus, for tubular members the following coefficients have been adopted: drag coefficient $C_D = 0.6$, inertia coefficient $C_M = 2.0$.

For the equivalent tubular leg with a diameter $D_e = 1.97$ m, $C_{De} = 2.376$ and $C_{Me} = 2.00$ have been derived.

It is observed from measurements in irregular waves that the drag coefficient stipulated above may be too low (e.g. see Ref. 5.). On the other hand, the inertia coefficient may be too large. Therefore, and in accordance with practice for stochastic analysis methods, the drag and inertia coefficients have been adjusted for one of the stochastic time-domain analyses carried out. The adjusted coefficients applicable to tubular members which have been chosen for this analysis are $C_D = 0.9$

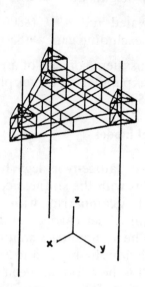

Fig. 1. Computer model.

TABLE 1
Calculated Natural Periods

Mode	Natural period T_0 (s)
First-mode sway, x direction	7·35
First-mode sway, y direction	7·41
First-mode torsion, z direction	5·78

and $C_M = 1.6$. For the equivalent tubular leg this corresponds to a $C_{De} = 3.25$ and $C_{Me} = 1.60$.

The other stochastic time-domain analyses were performed based on the regular-wave coefficients given above.

2.4 Natural periods

An eigenvalue analysis of the platform model shown in Fig. 1 was performed by SESAM.[2] The natural periods for the first modes are summarised in Table 1.

All higher-order modes are found to have natural periods below 1 s.

2.5 Damping

The total damping consists of structural damping, soil damping, and hydrodynamic damping.

All dynamic analyses of the platform have been based on the relative motion formulation of Morison's equation, and the hydrodynamic damping is thus inherent in the results presented.

The structural and soil damping has been included by the proportional Rayleigh damping model.[6] A value of 2% of critical damping for the period range around the fundamental period has been used.[1]

2.6 Environmental conditions

The storm environmental conditions considered in the analysis are

— maximum still-water level	75·0 m
— surface current	1·0 m/s
— current at −50 m	0·5 m/s
— current at seabed (current linear variation with depth)	0·5 m/s
— regular storm wave	
wave height	24·0 m
associated period	15·7 s

—irregular storm condition

significant wave height	12·9 m
zero up-crossing period	11·7 s
wave-energy spectrum	Pierson–Moskowitz[7]

Any short-crestedness, i.e. the angular distribution of the wave energy, has been neglected in the analysis of the extreme condition. The wave approach direction is along the positive x axis. Current is assumed to act in the same direction.

3 METHODS OF ANALYSIS

3.1 Wave theories

For slender structures (wave lengths greater than approximately five times the characteristic member dimension) influence of the structure on the fluid motion is disregarded. In such cases, wave kinematics are derived based on the undisturbed condition of the waves.

Two finite depth wave theories have been considered for the regular-wave analysis, these being the Stokes fifth-order wave theory[8] and the linear Airy wave theory.[9]

The original Airy wave theory only included wave loading up to the still-water level. Several methods of correction to include the wave kinematics in the crest have been proposed. One often-used correction is to specify the particle motions in the wave crest equal to the particle motions at the still-water level. This method is referred to as the Airy-SWL wave theory. A second method, proposed by Wheeler,[10] suggests that the particle motions are to be stretched or compressed according to the free surface. This modification is accomplished by using the water depth plus the surface elevation in the depth decay function instead of the constant water depth. These methods are illustrated in Fig. 2.

For the purpose of comparison, the total base shear for a single leg of the platform subjected to one cycle of a regular wave (H = 24 m, $T = 15·7$ s, no current) is also shown in Fig. 2, computed by FENRIS.[3] The maximum wave forces produced by this example show considerable deviations. At time instants when the leg is exposed to the full effect of the wave crest ($T = 4·0$ s), the Airy–Wheeler wave theory produces forces which are significantly smaller than the forces produced by the other two methods. However, the wave-force range obtained for one wave cycle is comparable for the three methods.

The figure illustrates the importance of selecting an appropriate wave

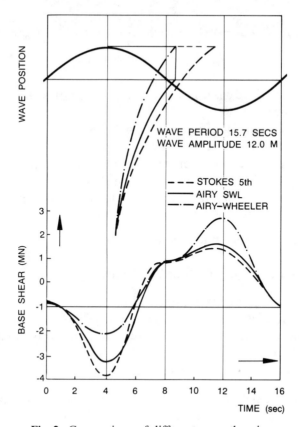

Fig. 2. Comparison of different wave theories.

theory for the problem being considered. Det norske Veritas recommends the use of the Airy-SWL wave theory, Stokes fifth-order wave theory, or other appropriate nonlinear wave theories for extreme analysis. The Airy–Wheeler wave theory is not recommended for this purpose and is not considered.

When irregular (stochastic) wave conditions are considered, superposition of wave components is normally used: this requires the use of a linear wave theory. The Airy-SWL wave theory has thus been used for deriving wave kinematics in irregular waves.

3.2 Wave forces

Wave forces are calculated by the use of the modified Morison's equation where motion components of the structure are included. For a circular member, this equation can be written

$$F = \frac{1}{2}\rho C_D D |u - \dot{r}|(u - \dot{r}) + \frac{\pi D^2}{4}\rho\dot{u} + \frac{\pi D^2}{4}\rho C_m(\dot{u} - \ddot{r}) \qquad (1)$$

where

F = total hydrodynamic force per unit length
ρ = density of water
C_m = added mass coefficient ($C_M = C_m + 1$)
u = Water particle velocity normal to the member
$\dot{u}$ = water particle acceleration normal to the member
$\dot{r}$ = structural velocity normal to the member
$\ddot{r}$ = structural acceleration normal to the member

3.3 Stochastic nonlinear time-domain analyses

These methods normally include the following steps:

— specification of the extreme condition;
— simulation of waves and current for a given sea-state;
— integration of the generalised nonlinear equations of motion in the time-domain;
— statistical processing of response histories.

The short-term sea-state is described by a wave-energy spectrum.

The time series of the water surface elevation is broken into sample points with constant sampling intervals Δt. Thus, waves with periods down to $2\Delta t$ are included in the simulation (cut off period). In all the time simulations reported in this paper a sampling interval $\Delta t = 1.5$ s has been used.

The equations of motion are solved in the time-domain by the Newmark-β method.[11] The time-step used in the time integration is 0.4 s, which corresponds approximately to 1/20 of the natural period of the platform.

As previously mentioned, the response of a jack-up platform subjected to wave and current loading is significantly nonlinear, due to such effects as

— the drag term in Morison's equation;
— wave surface effects;
— relative velocities between structure and fluid;
— geometric nonlinearities.

It is, therefore, obvious that the Rayleigh distribution is not suitable as a fit to simulated data and for the estimation of extreme values. For the simulated results derived here, a three-parameter Weibull distribution[12] is found to fit well. This distribution is described by

$$Q(x) = \exp\left[-\left|\frac{x - \gamma}{\beta - \gamma}\right|^{\alpha}\right] \qquad (2)$$

where

$Q(x)$ = probability of exceedance
x = stochastic variable (response amplitude)
α, β, γ = Weibull parameters

Generally, various distributions should be examined in order to obtain a best-possible fit.

Due to limited computer capacity, it is not practical to simulate continuously a storm of 3 or 6 h duration. Therefore, extrapolations of the simulated response to a storm of the required length have to be carried out. Typically, the duration of a time-simulation may be in the order of 20 min. Extrapolated results from one such 20-min simulation may thus be of limited value.

In the present study, one 40-min simulation and two different 20-min simulations have been carried out. In this way, different realisations of the storm are obtained and the basis for extrapolation is improved.

Additionally, two more 20-min simulations have been carried out based on the same simulated wave time series, but with other parameters changed. Thus, changes in the results of the parametric studies can be attributed to the parameters themselves.

The computer effort involved in one 20-min simulation of the considered jack-up platform is approximately 12–18 h CPU for a Vax Station 3100 (Digital Equipment Corp., Boston, US) (dependent on number of iterations to obtain equilibrium).

3.4 Regular-wave analyses

Regular-wave analyses for the purpose of design are normally carried out as static analyses. Maximum response for one design situation is obtained by stepping a complete wave cycle through the structure. As a design tool these methods are very simple, however, as previously discussed important effects are ignored.

Simple modifications to account for the most important of the ignored effects can, however, be carried out in an approximate manner as suggested in Ref. 1. These modifications are as follows.

Dynamic amplification

Inclusion of a dynamic amplification factor, DAF_{reg}, derived on the basis of a single-degree-of-freedom system.

$$DAF_{reg} = \frac{1}{\sqrt{\left(1 - \left(\frac{T_0}{T}\right)^2\right)^2 + \left(2\xi\frac{T_0}{T}\right)^2}} \tag{3}$$

where

T_0 = natural period
T = period of regular wave
ξ = damping ratio

Leg stiffness reduction

Geometric reduction in stiffness due to the presence of axial leg load

$$k_e = (1 - P/P_E)k \tag{4}$$

where

k = transverse overall stiffness of the platform
P = average force in one leg due to functional loads
P_E = Euler load for one leg

Δ-P effects

Second-order geometric effects due to lateral displacements (Δ) of the platform.

$$M' = n \cdot P \cdot \Delta \tag{5}$$

where

M' = total increase in global overturning moment
n = number of legs
Δ = lateral displacement of platform centre of gravity due to environmental loading

4 RESULTS OF STOCHASTIC ANALYSES

The dynamic response of the platform has been analysed in the time domain by the use of FENRIS,[3] taking into account nonlinear geometric and nonlinear hydrodynamic load effects. The nonlinearities in the loading include quadratic relative velocities in the drag term and integration of wave forces up to the instantaneous water surface. The structural modelling and boundary considerations of the jack-up platform are described in Section 2.

The responses have been analysed for an irregular sea state with a significant wave height of 12·9 m and a zero up-crossing period of 11·7 s (Pierson–Moskowitz spectrum). A current of 1·0 m/s at the surface (see Section 2) has been included. Wind forces on the deck structure are disregarded. Usually, wind forces are treated in a quasistatic way and will therefore have little influence on the dynamic platform responses.

Three separate time simulations of this sea state were conducted. In two of the runs the stochastic platform responses have been simulated for 20 min, and in one case for 40 min. A typical spectrum of the irregular waves being simulated is shown in Fig. 3. The generated waves closely follow the wave energy distribution of the Pierson–Moskowitz spectrum.

The results of the stochastic analysis of the dynamic platform responses from the 40-min simulation are described in Section 4.1. The dynamic responses are compared with similar quasistatic results established from one of the 20-min simulations after having increased the platform stiffness by a factor of 1000. The variations of the extreme dynamic responses derived from the 40-min and the two 20-min simulations are described in Section 4.2. Finally, the effect of altering the hydrodynamic properties (see Section 2.3) has been studied: these results are described in Section 4.3.

The presentation of platform responses is considered with reference to three characteristic design parameters. These are the vertical and horizontal seabed reactions and the leg-bending moment at the lower guides. All characteristic responses are calculated for the most highly

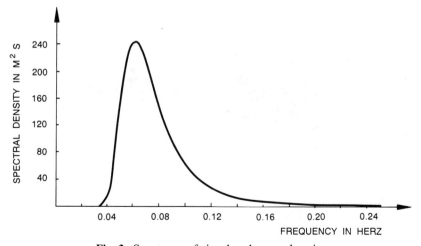

Fig. 3. Spectrum of simulated wave theories.

loaded leg (leeward leg). The units of the responses and the time scale of the plots presented in Section 4 are Newton, metre, and second, unless otherwise specified.

4.1 Comparison of dynamic and quasistatic platform responses in irregular waves

A typical time history of the dynamic platform responses (leg-bending moment) for the 40-min simulation is shown in Fig. 4. The response is offset (negative mean value) mainly due to the presence of current. The negative response amplitudes are somewhat larger than the positive amplitudes.

The associated response spectra of the vertical reaction (axial load), horizontal reaction (shear forces), and bending moment of the platform leg are shown in Figs 5–7. All the response spectra have two distinct peaks; one peak (0·06 Hz) corresponding to the peak in the wave spectrum (see Fig. 3), and one peak occurring at the natural frequency of the platform (0·136 Hz). The area of the resonance peak in relation to the total area of the spectrum reflects the dynamic contribution to the irregular responses. This contribution is significant for all the considered platform responses.

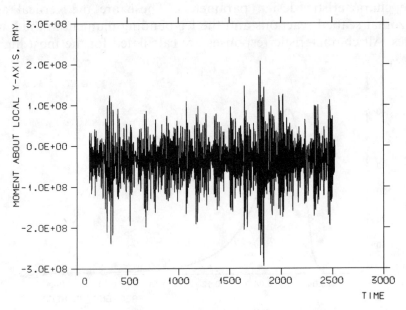

Fig. 4. Time history of leg-bending moment.

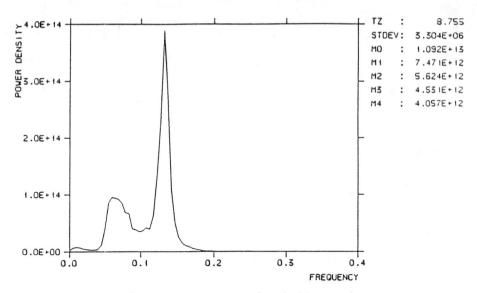

Fig. 5. Response spectrum of vertical leg reaction.

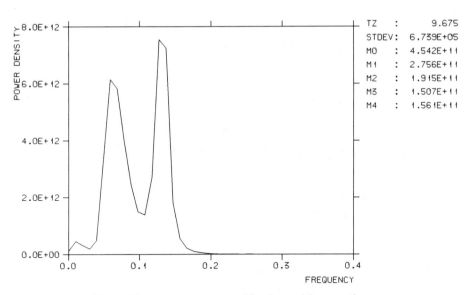

Fig. 6. Response spectrum of horizontal leg reaction.

A combined dynamic amplification factor for the response in irregular seas may be derived from the ratio between the standard deviation of the response spectrum and the standard deviation of the corresponding spectrum when disregarding the area under the peak at the natural frequency of the platform (0·136 Hz).

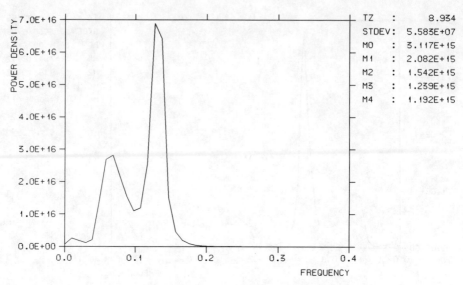

Fig. 7. Response spectrum of leg-bending moment.

$$\text{DAF}_{\text{irr}} = \sqrt{\frac{M_0}{M_0 - A_{\text{res}}}} \qquad (6)$$

where

M_0 = area of response spectrum
A_{res} = area of spectrum peak at natural frequency of platform

In this manner the following dynamic amplification factors have been derived:

vertical leg reaction $\qquad\qquad$ $\text{DAF}_{\text{irr}} = 1\cdot60$
horizontal leg reaction $\qquad\quad$ $\text{DAF}_{\text{irr}} = 1\cdot35$
leg bending moment $\qquad\qquad$ $\text{DAF}_{\text{irr}} = 1\cdot50$

The above amplification factors are difficult to apply in a meaningful way in other jack-up analyses or for other sea states. They consist of contributions from a range of wave frequencies and are closely interrelated to the nonlinear effects in the drag forces (relative velocities). It should also be remembered that the dynamic amplification factors for the considered responses will vary from one point on the platform to another, i.e. values will differ at a different leg. However, the estimated values do indicate the relative importance of the dynamics in the responses.

The time series of the responses have been treated statistically to establish histograms of the response amplitudes based on distributions of local maxima and minima of the signals. A three-parameter Weibull distribution has been fitted to the histograms and extrapolated to determine the most probable largest maximum and minimum values of the platform responses during a time period of 6 h. Examples of these statistical analyses for the positive and negative leg-bending moments are shown in Figs 8 and 9. Also included in these figures are the theoretical distributions fitted to the simulated responses. The associated characteristic parameters of these distributions are also given together with the most probable largest 6-hour values (extreme value).

The three-parameter Weibull distribution is defined by eqn (2) in Section 3. This distribution corresponds to the Rayleigh distribution for $\alpha = 2$, $\beta = \sqrt{2 \cdot \sigma}$, $\gamma = 0$, and to the exponential distribution for $\alpha = 1$, $\gamma = 0$, (σ = standard deviation).

With reference to the leg-bending moments in Figs 8 and 9, the values of the parameter α have been estimated as 1·57 for the positive amplitudes and 1·47 for the negative amplitudes. This implies a Weibull distribution of bending amplitudes in between the Rayleigh and exponential distributions.

In theory, the maximum values (absolute values) of the quasistatic wave and current loading on the platform will be dominated by the contribution from the drag forces. These maxima amplitudes are

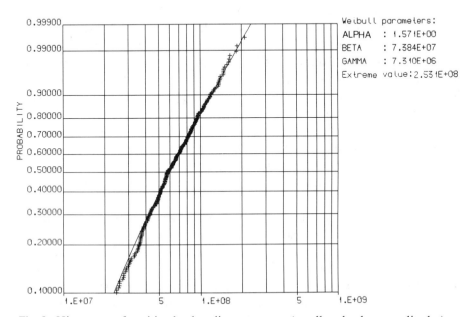

Fig. 8. Histogram of positive leg-bending moments (smaller absolute amplitudes).

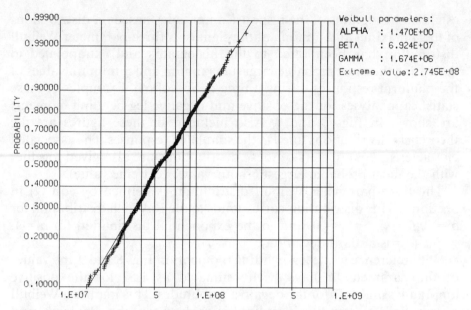

Fig. 9. Histogram of negative leg-bending moments (larger absolute amplitudes).

therefore expected to follow an exponential distribution. In order to study the actual variation of the nonlinear hydrodynamic loads in irregular seas, the platform responses of a rigid jack-up structure (*E*-modulus multiplied by a factor of 1000) have been analysed in the time domain. In this manner both the nonlinear geometrical and dynamic-response effects are excluded. The resulting time history, spectrum, and distribution of response amplitudes for the leg-bending moment (quasistatic wave-bending moment on the leg) are shown in Figs 10–13.

Comparison of the time histories in Figs 4 and 10 clearly shows the large difference between the dynamic and quasistatic leg-bending moments in irregular seas. The quasistatic time history contains less of the higher-frequency oscillations. Further, the positive amplitudes about the mean value are significantly smaller than those of the dynamic case.

The quasistatic-response spectrum is shown in Fig. 11. From this figure, it is seen that the resonance peak at the natural frequency of the platform (0·136 Hz) is no longer present, in contrast to the dynamic leg-moment responses shown in Fig. 7.

The histograms and distributions of the positive and negative bending moments are shown in Figs 12 and 13. The extreme value of the positive bending moments (see Fig. 12) is drastically reduced compared with the

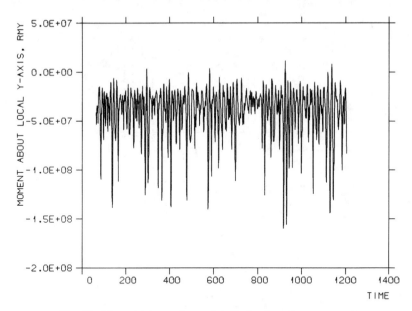

Fig. 10. Time history of quasistatic leg-bending moment.

dynamic-response value in Fig. 8. The estimated Weibull parameter ($\alpha = 1{\cdot}69$) for the quasistatic-moment distribution indicates that the inertia and drag contributions to the positive bending moments are of equal magnitudes. The same results have been found from static regular-wave analyses.

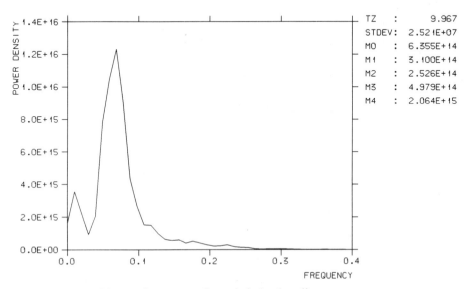

Fig. 11. Spectrum of quasistic leg-bending moment.

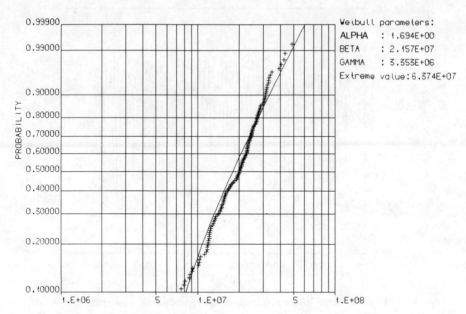

Fig. 12. Histogram of positive quasistatic leg moments (smaller absolute amplitudes).

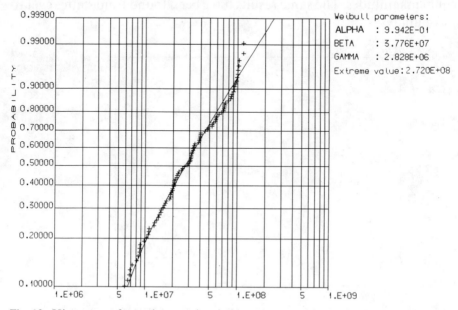

Fig. 13. Histogram of negative quasistatic leg moments (larger absolute amplitudes).

The negative bending moments are shown in Fig. 13. The α-parameter of this distribution is close to 1·0, which indicates that these amplitudes are dominated by drag forces and hence exponentially distributed. The extreme value in this case is practically identical to that of the dynamic-response case (see Fig. 9).

Based on comparison of results of the dynamic and quasistatic analyses in irregular waves, it can be concluded that dynamic effects have a great influence on the variation and magnitude of the resulting platform responses. On the other hand, the dynamic platform motions will have a damping effect on the nonlinear drag forces, and thereby reduce the apparent amplification of the platform responses.

The combined effect of wave and current will result in an offset of the responses both for the dynamic and quasistatic cases. Further, the extreme value and distribution of the positive-response amplitudes will differ from those of the negative amplitudes, but to a lesser extent when dynamic effects are included.

4.2 Variations in extrapolated extreme platform responses

The uncertainty and variation in extrapolated extreme values, depending on the simulation length of irregular seas, is a subject for considerable discussion. To investigate these effects, two additional time-domain analyses of the same sea state have been made. In these cases, the extreme platform responses have been derived from simulations of 20 min. The results are shown in Table 2, together with the extreme values of the 40-min analysis.

The given extreme responses represent the most probable largest values in a 6 h storm. Based on the theoretical Weibull parameters fitted to the results, the standard deviations of the extreme responses have been calculated.[12] For the considered cases, the variations of the extreme values are within 1·5 times the standard deviations of the corresponding responses. With reference to absolute values, a variation of approximately 15% is found.

4.3 Effect of altered hydrodynamic properties

The results presented in Section 4.1 and 4.2 are based on hydrodynamic properties applicable to regular-wave analysis (see Section 2.3).

With the altered hydrodynamic properties for stochastic analysis proposed in Section 2.3, an additional 20-min simulation has been performed. The alteration of properties corresponds to an increase of the

TABLE 2
Extreme Values Derived from 40- and 20-Min Simulations

Leg reaction	Amplitude value	Case 1 (40-min simulation)		Case 2 (20-min simulation)		Case 3 (20-min simulation)	
		Extreme value	Standard deviation	Extreme value	Standard deviation	Extreme value	Standard deviation
Vertical seabed reaction (MN)	max	54·20	2·04	NA	NA	50·77	1·65
	min	21·54	1·48	NA	NA	23·74	1·31
Horizontal seabed reaction (MN)	max	2·65	0·32	NA	NA	2·45	0·35
	min	−4·10	0·44	NA	NA	−3·58	0·40
Moment lower guides (MN × m)	max	214·5	24·8	218·7	26·0	180·8	22·8
	min	−313·1	29·6	−302·2	27·3	−277·0	28·1

NA — Not available

TABLE 3
Variation of Extreme Values with Hydrodynamic Properties

Leg reaction	Amplitude value	Hydrodynamic properties	
		$C_{De} = 2·376$ $C_M = 2·000$	$C_{De} = 3·249$ $C_M = 1·600$
Vertical seabed reaction (MN)	max	50·77	56·80
	min	23·74	22·61
Horizontal seabed reaction (MN)	max	2·45	2·50
	min	−3·58	−4·92
Moment lower guides (MN × m)	max	180·8	188·8
	min	−277·0	−349·6

equivalent drag coefficient for the jack-up legs by a factor of 1·38, and a reduction of the inertia coefficient by a factor of 0·8.

Comparison of the extreme responses for the two cases is shown in Table 3. The changes of the drag and inertia coefficients increase the positive amplitudes of the vertical and horizontal leg reactions, the negative amplitudes of the leg-bending moment by 30–40%, and the corresponding amplitudes of opposite sign by about 5%.

5 RESULTS OF DETERMINISTIC (REGULAR WAVE) ANALYSES

5.1 Computer analysis results

The following regular-wave analyses were performed by FENRIS:[3]

— static, linear analysis
— static, nonlinear analysis
— dynamic, nonlinear analysis

Stokes fifth-order wave theory, as well as Airy-SWL wave theory, were applied. The typical responses reported are

— vertical seabed reaction, leeward leg
— horizontal seabed reaction, leeward leg
— leg moment at lower guide, leeward leg

These characteristic responses are summarised in Table 4 when Stokes fifth-order wave theory is applied. Table 5 shows the same responses when the Airy-SWL wave theory is applied.

TABLE 4
Leg Responses, Regular-Wave Analyses Employing Stokes Fifth-Order Wave Theory
(Wave Period $T = 15.7$ s)

Leg reaction	Typical values	Static linear	Static nonlinear	Dynamic nonlinear
Vertical	max	52·00	54·61	70·77
seabed	min	33·40	32·97	11·46
reaction (MN)	range	18·60	21·64	59·31
Horizontal	max	0·63	0·66	4·70
seabed	min	−4·64	−4·37	−6·69
reaction (MN)	range	5·27	5·03	11·39
Moment	max	19·20	27·12	367·90
lower guide	min	−316·00	−359·20	−579·70
(MN × m)	range	335·20	386·32	947·60

TABLE 5
Leg Responses, Regular-Wave Analyses Employing Airy-SWL Wave Theory (Wave
Period $T = 15.7$ s)

Leg reaction	Typical values	Static linear	Static nonlinear	Dynamic nonlinear
Vertical	max	49·10	51·58	63·36
seabed	min	32·97	32·45	17·09
reaction (MN)	range	16·13	19·13	46·27
Horizontal	max	0·80	0·79	3·62
seabed	min	−3·90	−3·94	−5·73
reaction (MN)	range	4·70	4·73	9·35
Moment	max	28·36	37·50	288·40
lower guide	min	−268·10	−304·70	−473·90
(MN × m)	range	296·46	342·20	762·30

From this example it is seen that Stokes fifth-order wave theory predicts responses which typically lie 10–20% above the results of the Airy-SWL wave theory.

The nonlinear geometric effects in this system are moderate with a typical increase of the maximum vertical seabed reaction in the order of 5%. An increase in the lower-guide moment of approximately 12% is obtained. The small changes in the horizontal seabed reactions are due to the change in the internal force distribution within the three legs due to nonlinear geometric effects.

The dynamic effects which appear in this system when subjected to a periodic excitation are considerable. The amplification of the periodic wave response varies between 2·24 and 2·74 (dependent on type of response) when the excitation is predicted by Stokes fifth-order wave theory. The amplifications of the responses are reduced to between 1·97 and 2·42 when Airy-SWL wave theory is used.

Such high amplifications were not expected to be found in the dynamic regular-wave analysis, due to the rather high wave period compared with the first natural period of the platform ($T_0/T = 0.468$). However, when considering the response history which is shown in Fig. 14, it is seen that the higher-order frequency components in the drag loading, which are close to the first natural period of the platform ($T_0 = 7.35$ s), are present. These amplified higher-order components may explain the unexpected high amplifications in the results.

5.2 Results of simplified design approach

A simplified design approach for jack-up platforms, which is based on regular-wave analysis, is given in Ref. 1. Important effects are accounted for in a simplified manner, as given by eqns (3)–(5).

The dynamic amplification factor defined by eqn (3) is to be applied to the amplitude value of all wave responses. For the purpose of design, a range of wave periods has to be investigated in order to obtain maximum amplified responses. The following range of wave periods have therefore been considered;

T (wave)	DAF_{reg}
13·5 s	1·43
15·7 s	1·28
18·0 s	1·20

The second-order geometric effects defined by eqn (4) amplify the total leg-bending moment by a factor $1/(1 - P/P_E)$. For the example platform, this factor is 1·228.

Second-order geometric effects due to lateral displacement of the hull mass defined by eqn (5) result in an increased overturning moment. This moment will have to be balanced by additional vertical leg reactions. The increased overturning moment will also be amplified, due to the leg stiffness reduction defined by eqn (4). For the example platform, the increase in maximum overturning moment is calculated to be $M' = 113.4$ MN · $1.228 = 139.2$ MN (Stokes fifth-order wave theory).

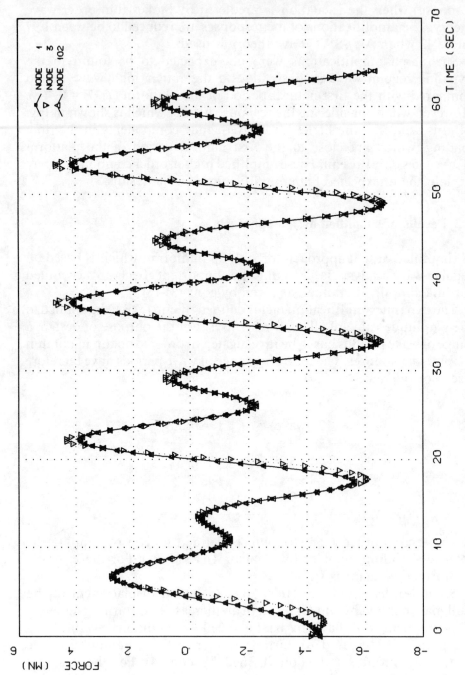

Fig. 14. Response history for horizontal seabed reaction, based on dynamic, nonlinear, regular-wave analysis.

The leg reactions obtained from the static linear analysis with regular waves (see Tables 4 and 5) have been modified according to the simplified procedure outlined above. The corresponding results are given in Tables 6 and 7.

From Tables 6 and 7 the following typical results can be deduced.

— The change in design wave period from 13·5 to 18·0 s causes only a moderate change in the predicted leg responses. Characteristic design values for vertical seabed reactions and leg-bending moment at the lower guide increase for reduced wave period. The horizontal seabed reaction, however, increases with increasing wave period.

— The change in wave period has two main opposing effects, namely

— the global wave loading and resulting leg reactions tend to increase with increasing wave period;
— the dynamic amplification increases with reduced wave period.

Generally, the sum of these effects is not easily quantified and analysis of different wave periods has to be carried out in order to find design values for the different response parameters.

— Second-order geometric effects as calculated in a simplified manner by eqns (4) and (5) correspond well with the full nonlinear static results derived by FENRIS (see Tables 7 and 5, respectively). However, second-order geometric effects do not seem to play an important role in this system when considering static responses.

TABLE 6

Leg Responses, Regular-Wave Analysis, Stokes Fifth-Order Wave Theory. (Simple Corrections for Dynamic Effects, eqn (3))

Leg reaction	Typical values	Corrected results[a]		
		(T = 13·5 s)	*(T = 15·7 s)*	*(T = 18·05 s)*
Vertical	max	58·78	57·64	58·00
seabed	min	28·31	29·94	30·90
reaction (MN)	range	30·47	27·70	27·10
Horizontal	max	1·67	1·36	1·25
seabed	min	−4·99	−5·07	−5·37
reaction (MN)	range	6·66	6·43	6·62
Moment	max	110·00	81·24	66·00
lower guide	min	−434·70	−413·00	−417·38
(MN × m)	range	544·70	494·24	483·38

[a] See Ref. 1.

TABLE 7

Leg Responses, Regular-Wave Analysis, Airy-SWL Wave Theory with $T = 15.7$ s. (Simple Corrections For Second Order Effects, eqns (4) and (5), and Dynamic Effects, eqn (3))

| Leg reaction | Typical values | Static linear | Corrected results[a] | |
			Second-order included	Second-order and DAF_{reg} included
Vertical	max	49·10	51·42	54·07
seabed	min	32·97	32·50	29·85
reaction (MN)	range	16·13	18·92	24·22
Horizontal	max	0·80	0·80	1·46
seabed	min	−3·90	−3·90	−4·55
reaction (MN)	range	4·70	4·70	6·01
Moment	max	28·36	34·83	86·04
lower guide	min	−268·10	−329·20	−382·18
(MN × m)	range	296·46	365·80	468·22

[a] See Ref. 1.

— Dynamic amplification obtained on the basis of eqn (3) deviates considerably from the full dynamic, nonlinear, regular-wave analysis obtained by FENRIS (see Tables 7 and 5, respectively). As discussed in Section 5.1, the considerable amplification which was obtained in the dynamic, regular-wave analysis may be due to amplified, higher-order frequency components of drag loading. Such effects cannot, of course, be predicted by eqn (3).

6 COMPARISON OF SIMPLIFIED AND STOCHASTIC APPROACHES

A comparison of typical jack-up responses calculated by the simplified design approach based on the regular-wave analysis and results of the stochastic nonlinear time-domain analysis is presented in Table 8. The results of the stochastic analysis are based on the 40-min time simulation extrapolated to a 3-h storm. The reason for selecting a 3-h storm for the comparison is that the characteristic largest wave amplitude in this storm is equal to the design wave height used in the regular-wave analysis.

Furthermore, the extrapolated results of this time series have been corrected to include more appropriate hydrodynamic coefficients for stochastic analysis as given in Section 2.3.

From Table 8, it can be seen that the characteristic design leg reactions as obtained by the regular-wave analysis (Stokes fifth-order wave theory) with applied simple second-order and dynamic corrections are within 0·97–1·12 of the corresponding stochastic results. If Airy-SWL wave theory is used as the basis for the regular-wave analysis, the characteristic design leg reactions are within 0·87–1·03 of the same stochastic results. It should be noted that the variations of these results fall within the expected variation of the stochastic results themselves (see Section 4.2). If no corrections for second-order and dynamic effects were employed (see Tables 4 and 5), comparisons with stochastic results of characteristic design leg reactions give ratios of 0·85–0·88 (Stokes fifth-order wave theory) and 0·74–0·80 (Airy-SWL wave theory).

A clear trend from Table 8 is the wider range (range = max–min of same characteristic leg response) in the results of the stochastic analysis as compared to the regular-wave analysis. It is also seen that the vertical and horizontal seabed reactions (characteristic design values), as predicted by the regular-wave analysis, are somewhat smaller than the stochastic results. However, leg-bending moments (characteristic design values) are somewhat higher.

The above results indicate that the regular-wave analysis corrected for second-order and dynamic effects seem to predict characteristic design values reasonably well. However, such methods cannot reproduce the total dynamic response of the structure in irregular seas.

7 CONCLUSIONS

The extreme wave and current response of a typical North Sea jack-up drilling platform in 75 m water depth has been calculated using nonlinear, stochastic, time-domain analysis techniques. The results have been compared to the results of more simplified methods used in jack-up designs.

The main conclusions from these analyses and comparisons are summarized below. Although these results are strictly valid only for the example jack-up subjected to the given conditions, it is believed that the results illustrate phenomena which may be rather typical for this type of structure.

Results of stochastic analyses show that dynamic effects have a significant influence on the variation and magnitude of jack-up responses. For the example platform subjected to irregular waves,

TABLE 8

Comparison of Results from Simplified, Regular-Wave Analysis and Stochastic Nonlinear Time-Domain Analysis

Leg reaction	Typical values	Regular analysis (T = 15·7 s)		Results of stochastic analysis
		Airy-SWL wave theory	Stokes fifth-order	
Vertical seabed reaction (MN)	max	54·07	57·64	59·52
	min	29·85	29·94	20·84
	range	24·22	27·70	38·68
Horizontal seabed reaction (MN)	max	1·46	1·36	2·62
	min	−4·55	−5·07	−5·24
	range	6·01	6·43	7·86
Moment lower guide (MN × m)	max	86·04	81·24	216·34
	min	−382·18	−413·00	−369·90
	range	468·22	494·24	586·24

dynamic amplification factors of 1·60, 1·35, and 1·50 have been estimated for the vertical and horizontal leg reactions and the leg-bending moment at the lower guides, respectively (most heavily loaded leg).

In regular waves, dynamic amplification factors of 2·00–2·75 have been found. Such high amplifications were not expected to be found for the regular-wave analysis, as the ratio between the natural platform period (T_0) and the regular-wave period (T) is $T_0/T = 0·468$. The main cause for these high dynamic amplifications in regular waves is found to be super-harmonic load effects. Such effects become less dominant in irregular waves. As a consequence, the responses obtained by dynamic analyses in regular waves may be significantly overpredicted.

The combined effect of waves and current will result in an offset of the responses. Under such conditions, the extreme value and distribution of the positive response amplitudes will differ somewhat from the extreme value and distribution of the negative amplitudes.

Based on results from three separate time simulations in irregular seas, the variations of the extrapolated 6-h extreme values are found to be within 1·5 times the standard deviations of the corresponding extreme responses. With reference to absolute values, this variation corresponds to approximately 15% of the extreme responses.

Comparison of results from regular(design)-wave analysis and full stochastic time-domain analysis indicates that the regular-wave analysis, when simple corrections for second-order and dynamic effects are applied, predicts characteristic design responses within the expected variation of the extreme values from different stochastic simulations. It should, however, be noted that these simplified methods cannot predict the full dynamic response of the structure in irregular waves. This is illustrated by the fact that the dynamic ranges of the characteristic platform responses were underestimated by approximately 25–35% by these simplified methods.

Second-order geometric effects, as calculated by simple correction factors, correspond well with the results of nonlinear finite-element analysis. However, second-order geometric effects are moderate for the case considered here.

When no corrections for second-order and dynamic effects are introduced, the results of the regular(design)-wave analysis underpredicts the extreme responses by approximately 15–25% compared to the stochastic results.

It should be observed that the hydrodynamic coefficients have been adjusted for the stochastic analyses when used for comparison with the regular-wave analysis, based on measured values in irregular seas. Further, the effect of short-crested seas has not been included.

REFERENCES

1. Det norske Veritas, Strength analysis of main structures of self-elevating units. Classification Note No. 31.5, Hoevik, Norway, May 1984.
2. SESAM (Super Element Structural Analysis Modules) Computer System, Det norske Veritas, Hoevik, Norway, 1989.
3. FENRIS (Finite Element Nonlinear Integrated System) Computer System, Det norske Veritas, Hoevik, Norway, 1989.
4. FENSEA User Manual (preliminary), Det norske Veritas, Hoevik, Norway, May 1988.
5. Department of Energy, Hydrodynamic and hydrostatic loading on rigid structures. Background document to Guidance on Design, Construction and Certification (Draft Fourth Edition, Rev. A), HMSO, London, October 1988.
6. Clough, R. W. & Penzien, J., *Dynamics of Structures.* McGraw-Hill Inc., New York, 1975.
7. Pierson, W. J. & Moskowitz, L., A Proposed Spectral Form for Fully Developed Wind Seas Based on the Similarity Theory of S. A. Kitaigorodskii. *Jour. of Geophys. Res.,* **69** (1964) 5181–90.
8. Skjelbreia, L. & Hendrickson, J., Fifth Order Gravity Wave Theory. In *Proc. 7th Conference on Coastal Eng.,* The Hague, 1960, pp. 184–96.
9. Lamb, H., *Hydrodynamics.* Cambridge University Press, Cambridge, 1932.
10. Wheeler, D. J., Methods for calculating forces produced by irregular waves. *Journal of Petroleum Technology,* **22** (1970) 359–67.
11. Bergan, P. & Mollestad, E., Static and Dynamic Solution Strategies in Nonlinear Analysis. In *Numerical Methods for Nonlinear Problems*, Vol. 2, ed. C. Taylor, E. Hinton & D. R. F. Owen. Pineridge Press, Swansea, 1984, pp. 3–17.
12. Gran, S., A course in ocean engineering. Technical Report 88–2001, A. S. Veritas Research, Oslo, 1988.

The Development of a Stochastic Non-linear and Dynamic Jack-Up Design and Analysis Method

C. J. Mommaas & W. W. Dedden

Marine Structure Consultants (MSC) BV, PO Box 687, 3100 AR Schiedam,
The Netherlands

ABSTRACT

This paper describes the development of a non-linear, dynamic jack-up analysis method in the time domain. It provides background as to why and when such analysis is required.

The theoretical background of the methods applied are discussed and the main features of the programme are described.

Key words: jack-ups, dynamics, non-linear response, time-domain analysis

1 INTRODUCTION

The *Jack-Up* is a self installing mobile offshore platform. Jack-ups operate therefore in several very different modes (Fig. 1). They are towed, with the legs fully elevated, from one location to another in a floating mode. In the installation mode the legs are lowered to the seabed, the hull is jacked up out of the water, the leg foundation is pre-loaded, and the hull is finally jacked up to the final operational airgap. In the elevated mode the jack-up provides its final platform function supporting various offshore operations. In most cases this will be drilling operations. Other applications are: accommodation, construction support and production.

This paper describes the development of an integral method to analyse jack-ups in the elevated mode. The purpose of such an analysis is to design or to verify the function of the jack-up in this mode, which is to

155

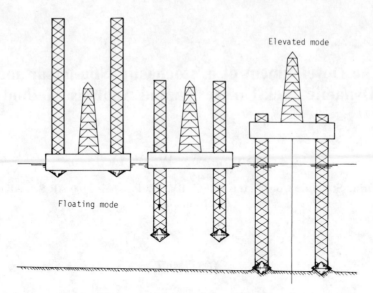

Fig. 1. Jack-up operational modes.

provide an economic but stable and safe platform supporting the weights and operational loads and resisting the environmental forces from wind, current and waves.

2 JACK-UP ANALYSIS METHODS

The analysis of a jack-up in the elevated mode involves many interrelated aspects:

— definition of functional loads;
— definition of environmental conditions;
— calculation of environmental loads;
— structural analysis of the overall jack-up structure;
— geotechnical analysis of the seabed foundation;
— evaluation of the analysis results with performance and safety criteria.

During the lifetime of a jack-up such an analysis is performed on many occasions by many different parties, for different purposes, as demonstrated in Table 1.

During the 30 years of existence of jack-ups, these parties have developed analysis tools suitable for their particular function within the industry. In present day practice the tools used for such analysis may vary from simple hand calculations to multi-hour computer calculations

TABLE 1
Overview of Jack-Up Analysis Activities

Jack-up analysis activity	Party/parties
Concept development[a] To find the optimum configuration/dimensions to suit the operational and economic requirements	Designer
Concept appraisal[a] To verify/certify that a particular concept design complies with specified requirements	Classification Society, warranty surveyor, consultants on behalf of (future) owners
Basic design[b] To completely document the final dimensions and scantlings with verification against requirements from classification society, national authorities, etc.	Designer
Classification[b, c] Verification of the basic design by independent analysis for each unit built	Classification Society
General appraisal of the jack-up[a] To verify the suitability of the jack-up for the operations and operational area envisaged when planning a (long term) contract	(Future) operator or a consultant on his behalf
Site assessment[a, c] To check the suitability of the jack-up for use on a particular location	Warranty surveyor, owner, operator, consultant on behalf of owner or operator, designer on behalf of owner
Modification or conversion[c] Re-analysis in concept or of basic design to support major modification or conversion	Designer or consultant

Note on the amount of detail in the analysis of the jack-up analysis activities:
[a] In terms of global parameters only, i.e. overall displacement, leg bending moments, leg reactions, general stress level etc.
[b] In full detail to provide documentation/verification on all dimensions and scantlings.
[c] In terms of global parameters and supported by the original design documentation to provide more detail.

depending on the purpose of the particular analysis. Any analysis method of a system as complex as a jack-up is based on a number of approximate theories, assumptions and simplifications. Approximate theories are used because better ones are not available or simply because they require less calculation time and resources than exact methods. Assumptions are being made either because the exact situation is unknown or to provide a shortcut in the analysis. Simplifications are necessary to reduce a structure consisting of several thousands of elements in a complex environment of wind, waves, current and operational loads to a manageable analytical problem which can be solved within a reasonable resource limit.

Together with an appropriate safety philosophy and safety factors, the combination of approximations, assumptions and simplifications can form a perfect analysis tool for a jack-up in a particular area of application. Presently available tools dedicated to jack-up analysis within the industry (each with their own set of approximations, assumptions and simplifications) are generally developed for the jack-up for exploratory drilling in water-depths up to approximately 70 m.

The last few years have shown a growing interest for the application of jack-ups not only in harsher environments and deeper waters but also for other purposes such as development drilling and production. Such applications are characterised by:

— more important environmental loading;
— more time on one location;
— more interest and influence from the operator (oil company).

To be prepared for this, the jack-up industry has been investigating isolated analysis issues in joint industry studies with the purpose of improving knowledge in some areas (dynamics, foundations, etc.).

More recently the view has been introduced that, together with improvements of isolated issues, a more complete review of the entire set of approximations, assumptions, simplifications and safety factors is required.[1-4] With this process in progress, the need was recognised for a new jack-up analysis tool that would be complete, consistent and better equipped to cover the present requirements on the one hand, and easily accessible, practical and cost effective on the other hand.

3 OBJECTIVES

At the outset of the development of the methods for this tool, the following objectives were formulated:

— The method must comprise not only of analysis but also input, model generation and output interpretation; in short it must result in a practical integral jack-up analysis tool.
— The method shall be suitable for use in concept design, design evaluation and as a pre-analysis tool for basic design. Input shall

not be in terms of coordinates, member incidents, etc., but in terms of an implicit jack-up structure, i.e. distance between leg centres, shape of leg, number of chords per leg, bracing system, etc. Also the output shall not be in terms of, for example, moment M_x at node 463 or force F_x at node 12, but as: leg bending moment at lower guide level and vertical leg footing reaction for leg nr 1.

— The method may of course contain simplifications but shall be theoretically complete and consistent. The method shall only use state-of-the-art methods.

— The method is to be aimed at the analysis of the overall performance of the jack-up, i.e. it provides overall displacements of the hull, bending moments in the leg, etc. and not the stress level in, for example, bracing nr 74 at node 372, point E.

— The method shall provide the results of the overall analysis in terms of (most probable) extreme values.

— The method shall result in a tool that can produce results within two days after availability of the required input data.

4 BACKGROUND TO METHODS

4.1 General

The method presently used can (and this is probably true for most methods used by the jack-up industry today) be characterised as follows:

Quasi-static:
A regular wave is stepped through the hydrodynamic model and calculation is continued for the forces found at the position of occurrence of the maximum overturning moment.

Semi-dynamic:
Dynamic effects are accounted for by a DAF calculated on the basis of a single mass spring system analogy.

Partly non-linear:
Non-linearity of the hydrodynamic forces and geometric non-linearity of the overall structure (P-δ) are included.

Deterministic:
Evaluation is based on the results of a maximum design wave.

Single direction:
All environmental action on and structural properties of the jack-up are assumed to be colinear thus reducing the analysis to a two-dimensional problem.

This method has been made into a practical design tool in the form of a computer program called MSC-LEGLOAD. Figure 2 provides an overview of the one page input/output of this program. The requirements set out for the development of the new method lead very quickly to the following main characteristics:

Dynamic effects:
The dynamic behaviour will become more important for jack-ups moving into deeper water using a DAF factor on the basis of the analogy that a single mass spring system is rather coarse and is claimed to be incorrect in some cases.[3]

Irregular waves:
Proper dynamic analysis requires consideration of the irregular nature of the environmental forces.

Fully non-linear:
The overall structural behaviour of the jack-up involves several non-linearities:

- leg to hull interface (Fig. 3)
- overall leg hull structure (Fig. 4)
  - P-δ effect
  - Euler effect
- leg to soil interface (Fig. 5)

The environmental force also involves non-linearities (Fig. 6):

- Morison drag force calculation
- wave elevation and crest kinematics
- combination current-wave

Multi-directional:
In reality, waves, current and wind are not necessarily colinear. Also the energy from waves and wind is not concentrated in one direction.
Solution in the time domain:
With the selection of above characteristics the only practical method of solution is time domain simulation.
Statistical treatment of results:
Time domain analysis results require statistical treatment to produce comprehensible results. Performance evaluation requires information on the extreme (most probable maxima) of various parameters.

With these characteristics in mind the details of the method were further developed on the basis of the following procedures:

- collection of methods available/described in literature
- discussion on their validity/feasibility/practicality for this project
- selection of method to be implemented

```
———————————————————————————————  *      * ****** ****** —   1
——                L E G L O A D      —— **  ** *        *     ——
                                       —— * **  * ****** *      ——
—— Version:  3.06.                     —— *      *      * *     ——
—— Apollo Domain       23/09/1989   13:31:09 —— *    * ****** ****** ——
—— MARINE STRUCTURE CONSULTANTS ——                    — HOLLAND ——
```

Unit: CJ46 R60 standard
Case: Survival condition

Leg No:	1	2	3
Coordinates =X=:	−13.300 m	26.700 m	−13.300 m
=Y=:	−23.000 m	0.000 m	23.000 m

Leg guide distance......:	1.000 m	Top jackh. above base...:	15.600 m
Leg inclination..............:	2.500 o/oo	Lower guide above base:	5.000 m
Leg specific weight......:	46.94 kN/m	Buoyancy ....:	18.34 kN/m
Total leg length............:	111.8 m	Moment of inertia........:	5.570 m4
Specific diameter.........:	10.000 m	CD: 0.366 CM:	0.052 —
Footing subm. weight..:	1.073 MN		
Windarea of hull.........:	1700.0 m2	Level above base...........:	16.000 m

*** Operational data ***		*** Environmental conditions ***	
Elevated weight...........:	86.33 MN	Waterdepth......................:	64.00 m
Eccentricity...... —X—:	−0.40 m	Wave height.....................:	23.00 m
—Y—:	−0.10 m	Wave period.....................:	15.60 sec
Wave angle.....................:	210.0 degr	Wave theory.....................:	Stokes-5
Airgap..............................:	15.48 m	Wave elevation................:	13.98 m
Clearance......................:	1.50 m	Wind velocity..................:	42.50 m/sec
Support position...........:	4.00 m	Current @ surface..........:	1.00 m/sec
Penetration....................:	7.60 m	Current @ bottom..........:	0.00 m/sec
Footing restraint...........:	0. MN.m/degr	(Type 0)	
Platform deflection.....:	0.96 m		
Natural period..............:	5.14 sec		
Overturning stability...:	1.91 over legs 1 and 3		
Wave/current force......:	9.47 MN	Overturning moment....:	514.6 MN.m
Wind force......................:	2.39 MN	Overturning moment....:	235.7 MN.m
Auxiliary forces ....:	0.00 MN	Overturning moment....:	0.0 MN.m

Forces in leg...........................:	1	2	3
Wave/current force.(MN):	3.116	2.547	3.805
—level.(m):	52.991	53.748	55.862
—angle.(d e g r):	210.000	210.000	210.000
deflection.(m):	0.726	0.730	0.755
moment.(MN.m):	238.257	227.740	249.402
Guides................hor.(MN):	0.671	1.505	0.212
..............vert.(MN):	47.752	9.559	29.019
......moment.(MN.m):	273.983	283.072	262.948
Footing...............hor.(MN):	3.786	4.052	4.017
..............vert.(MN):	52.899	14.706	34.166
......moment.(MN.m):	0.000	0.000	0.000

Fig. 2. Input/output for program MSC-LEGLOAD.

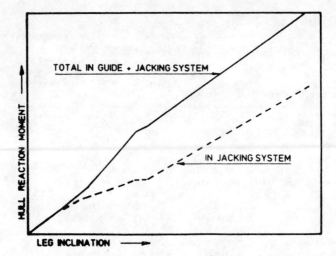

Fig. 3. Non-linearity of leg to hull interface (see Ref. 5).

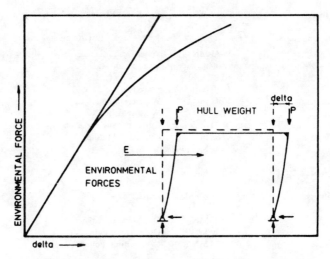

Fig. 4. Non-linearity of overall jack-up structure.

This process is further described for the three main sections of the method: hydrodynamics, structural and statistics.

4.2 Hydrodynamics

For the calculation of hydrodynamical forces on the jack-up, three main subjects can be distinguished:

—generation of wave kinematics

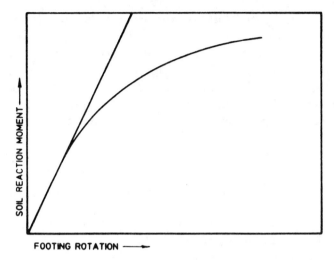

Fig. 5. Non-linearity of leg to soil interface (see Ref. 6).

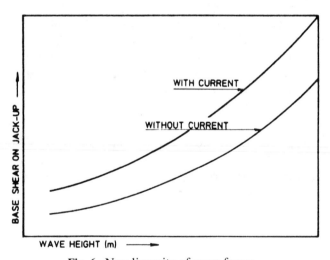

Fig. 6. Non-linearity of wave forces.

— hydrodynamic properties of the jack-up
— force calculation

The theoretical backgrounds of the methods as applied in the time domain program will be discussed in more detail.

Jack-up leg cross-sectional dimensions are small relative to the (design) wavelength. The disturbance of the wave field caused by the presence of the legs is therefore small and hydrodynamic forces on the legs can be calculated by applying the Morison equation.

For lattice legs, being built up from several elements (chords, braces, etc.), this would require assembling the leg hydrodynamic force from the loads on each structural element, taking into account the orientation, dimensions and hydrodynamic properties of each member and the proper wave/current kinematics along it. In the time domain the hydrodynamic loads will have to be recalculated for each time step, so the cumbersome wave/current force assembly has to be repeated many times.

In design waves, phase differences between the water motion at the various leg elements will be small. If they are neglected, the contributions of all leg members taking into account their orientation (relative to the flow), dimensions (length and diameter), shape (i.e., non-circular chords) and drag and inertia coefficients, can be summed to find the equivalent leg diameter, C_d and C_m values. This simplification is often referred to as the leg 'stick' model.

When applying leg stick models, the effect of vertical forces, arising from non-horizontal (wave) motion around more or less horizontal elements in the leg is also neglected. These vertical forces contribute to the overall overturning moment on the structure. However, comparing contributions from horizontal and vertical (wave) loads to the total overturning moment of the jack-up indicates that the effect of the vertical loads usually tends to be less than 5% of the effect of the horizontal loads. Neglecting vertical hydrodynamic forces seems justified therefore, saving a lot of extra (at each time step repeated) calculations.

However, application of the 'stick' model still requires knowledge of the C_d and C_m coefficients of each leg member. For round, cylindrical elements, a wealth of data are available, partly based on scale tests, partly on measurements on actual structures. However, it is well known that test data of C_d and C_m show a wide scatter.[7]

For smooth, round cylinders, the two main parameters that seem to influence the C_d and C_m values are:

Keulegan Carpenter number $\quad K_c = 2\pi\dfrac{A}{D}$

Reynolds number $\qquad\qquad\quad R_e = \dfrac{UD}{V}$

where

A = typical water motion amplitude,
U = typical water velocity amplitude,
D = member diameter,
V = kinematic viscosity of water.

At high Reynolds numbers (say above 10^5), corresponding to a post-critical flow regime, the C_d and C_m coefficients show a negligible dependency on the Reynolds number. For jack-ups legs in extreme wave conditions the Reynolds number will generally be above 10^5 (leg members are rough due to marine growth), so the Reynolds number dependency can be neglected.

To take into account the Keulegan Carpenter dependency at each time step during the simulation, the ratio A/D should be calculated and the hydrodynamic coefficients varied accordingly. In Ref. 8 a C_d and C_m Keulegan Carpenter number dependency is proposed, based on the comparisons with experimental and full scale measurements. The typical water motion amplitude A is taken as:

$$A = A_{\text{waves}} + \frac{T}{4} V_{\text{current}}$$

where

A_{waves} = typical horizontal wave motion amplitude,
T = typical wave period,
V_{current} = current velocity.

Note that in this way the typical water motion amplitude used in the calculation of the Keulegan Carpenter number is depth dependent but constant in time. At the expense of a small decrease in realism a CPU-consuming K_c number evaluation at each time step during the simulation can be avoided.

For most jack-ups, the legs cannot be seen as a collection of smooth circular cylinders. Instead, large excrescences like racks are present. Data on C_d and C_m for such shapes are not as abundant as for circular cylinders. Some data can be found in Ref. 9. Due to the limited data on C_d and C_m coefficients for non-circular shapes, a Reynolds and/or Keulegan Carpenter dependency is not very clear.

An important part of the program is the part that generates the wave kinematics. A number of methods are available for generation of the wave signals of a certain spectrum.

4.2.1 Summation of sines

$$\text{output}(t) = \sum_{i=1}^{N} a_i \cos(\omega_i t + \phi_i)$$

where

t = time,

ω_i = frequency,
a_i = determined from the spectral shape,
ϕ_i = random phase.

4.2.2 Filtering of white noise

$$\text{output}(t) = \int_{-\infty}^{\infty} h(\tau) R(t - \tau) \, d\tau$$

where

$R(t)$ = Gaussian white noise input,
$h(\tau)$ = impulse response function,
τ = time lag.

4.2.3 Auto regressive moving average (ARMA) models

$$\text{output}(t) = -\sum_{i=1}^{N} a_i V(t - i \times dt) + \sum_{k=0}^{M} b_k R(t - k \times dt)$$

where

a_i, b_k = ARMA coefficients,
V = output signal,
R = white noise,
dt = time step.

Of these three methods the first is the one applied most in offshore engineering. Its charm is its simplicity. White noise filtering and ARMA techniques are more complex and require careful handling.

A disadvantage of the sines summation method is its discrete nature: the spectrum energy is allocated to a discrete number of frequencies. The error introduced in this way is discussed in Ref. 10; a part of the random character of the signal would be lost. This can be avoided however by building up the kinematics time record from a number of subrecords. At the start of each subrecord the amplitude of each frequency component is drawn from a Gaussian distribution; in this way the amplitude of each wavelet is assumed to be a stochastic variable. On the other hand if the number of frequency components is taken to be sufficiently large the random character of the signal should be sufficiently guaranteed. In particular when an Inverse Fast Fourier Transform algorithm is applied the sine summation method forms a very efficient wave generation method. For these reasons and the limited experience with white noise filtering and ARMA techniques for generation of wave kinematics in the time domain program, the sines summation method has been adopted.

In reality, at sea, the wave energy will not only be distributed over the frequency but also over all the directions. Whereas for wave energy as a function of frequency, several types of standard spectra are proposed (Pierson–Moskowitz, Jonswap, etc.), for directionality cosine power law functions are often proposed. The program offers this possibility, but also has the option to define separate independent wave spectra for a number of wave directions. In this way for instance the combination of a swell spectrum and 'young' wind wave field can be simulated.

Modelling the irregular sea as a sum of sine waves (Airy waves) results in symmetrical properties of the wave profile. No distinction is made between crests and troughs, so their statistical properties are identical. Actual irregular waves do show an asymmetric behaviour with crests higher than troughs. In the literature some methods are proposed to introduce asymmetry.[11, 12] However, in light of the uncertainty regarding the correctness of these methods, and the lack of consensus in the offshore engineering industry on how to take into account this asymmetric behaviour, asymmetric waves have not been implemented in the program.

Another problem arising from the use of linear wave theory is the derivation of wave kinematics above the still water level. Linear first order theory does not yield velocities for positions above the MSL because it assumes the waves to be infinitely small. In Ref. 13 Rodenbusch and Forristall discuss three models for the determination of the velocities in the crests:

— stretching
— extrapolation
— multi-extrapolation

The same authors propose the use of a combination of stretching and extrapolation, called delta stretching, based partly on theoretical backgrounds and partly on evaluation of measured (model and full scale) wave kinematics data.

The delta stretching model (Fig. 7) is a general formulation for the adaptation of the total velocity profile, with pure stretching (surface velocity kept constant) and pure extrapolation (velocity at still water level kept constant) being special cases.

Besides waves, current can also be present. In general the current speed will be depth dependent and is usually sufficiently approximated by a form of power function. However, current and waves do influence each other. For instance the current profile is usually defined relative to mean sea level, and should therefore be adapted to the instantaneous

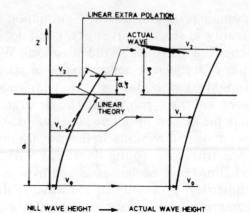

Fig. 7. Delta stretching.

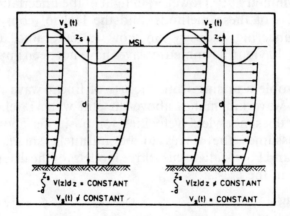

Fig. 8. Example of current stretching.

water level in the presence of waves. Several methods for stretching the current profile separately are available; Fig. 8 provides an example.

In the case of delta stretching, as discussed earlier, the total velocity profile (current and wave velocities combined) are stretched.

Wind loads on the hull are calculated from the specified wind areas. Wind forces on the leg parts above the jackhouse and between the sea level and the base of the hull are calculated by the program, using the leg ('stick') drag coefficients. Calculated wind loads are assumed to be stationary forces, so gustiness is not considered in that case. Non-stationary wind forces, or in fact any auxiliary non-stationary force, can be defined as time histories.

4.3 Structural

The structural representation of the jack-up has to fulfil the following requirements:

— suitable to provide meaningful results of overall jack-up design parameters such as displacement of the platform, leg bending at lower guide, foot reactions, etc., as a result of a multi-directional environmental loading and the weight to be supported.
— to provide the correct linear and non-linear stiffness representation of the hull, leg to hull connection, the legs and the leg to seabed connection
— to provide the correct mass distribution of major weights such as platform (elevated weight) and legs
— to be suitable for application in a time domain analysis, i.e. with a minimum amount of degrees of freedom, elements and masses

In this context it was decided to use a structural model as shown in Fig. 9. In this model the legs are represented by a variable number of beam elements. The elements are given equivalent properties for axial, bending and shear deformation. The stiffness representation of the leg element is such that second order effects (P-δ and Euler) are properly accounted for.

The leg masses are lumped at the nodes between the beam elements. The part of the leg projecting above the jackhouse level is structurally not included in the model. The mass of that part of the leg is included in the lumped mass at the hull to leg connection node. The hull is represented by a number of beams provided with bending and torsional stiffnesses to correctly represent the 'out of plane' stiffness of the hull. The hull mass is divided over the centre node and the leg connection nodes such that possible hull weight eccentricity and rotational inertia of the mass can be represented properly.

The stiffness of leg to hull connection is represented by two non-linear rotational springs. The actual connection of the leg to the hull is a complicated structure consisting of several mechanical systems in many different arrangements for the various types of jack-ups. Figure 10 shows a typical arrangement.

Transition forces between leg and hull are transferred by essentially two systems (see Fig. 11):

— the horizontal guiding system capable of transferring the bending moment by a set of horizontal forces;

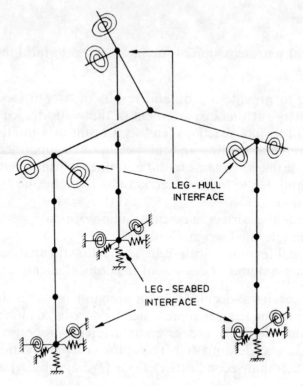

Fig. 9. Overall jack-up model.

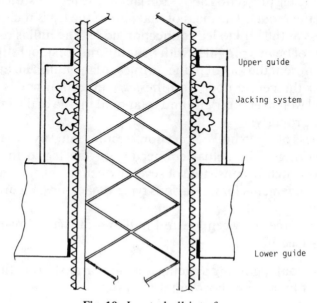

Fig. 10. Leg to hull interface.

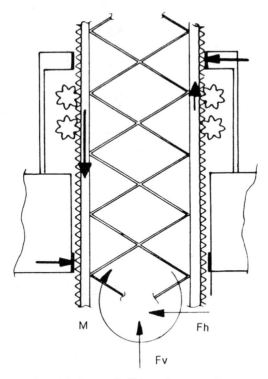

Fig. 11. Leg to hull interface reactions.

—the vertical jacking systems and/or fixation system capable of transferring both vertical forces and bending moments by a set of vertical forces.

The distribution of the leg bending moment over both systems is strongly dependent on their relative stiffnesses and clearances. Consequently the moment inclination relation of the springs in a structural model is determined by the many aspects determining stiffness and clearance in the entire leg to hull interface:

—bending, shear and torsional stiffnesses of the leg section between the guides;
—axial bending, shear and torsional stiffnesses of the jackhouse structure;
—stiffness of the guides;
—stiffness of the jacking system (pinions);
—amount of clearance of the leg within the guides (Fig. 12);
—amount of backlash in the jacking system (Fig. 13);

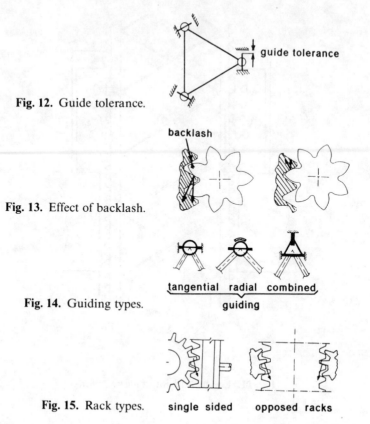

Fig. 12. Guide tolerance.

Fig. 13. Effect of backlash.

Fig. 14. Guiding types.

Fig. 15. Rack types.

—type of leg guide arrangement (radial, tangential or combined; Fig. 14);
—rack arrangement (single radial or double opposed; Fig. 15).

A dedicated non-linear FEM model has been developed to calculate the spring characteristics of the leg to hull interface.[5] The model used is shown in Fig. 16.

At the lower guide a leg inclination is imposed and the resulting moment is calculated by an incremental procedure. A sensitivity analysis with this method shows that the spring characteristics are:

—non-linear due to the guide clearances and backlash (see Fig. 17)
—dependent on the vertical reaction in the leg, through interaction with the backlash (see Fig. 18)
—dependent on the orientation of the imposed inclination (see Fig. 19)
—to be coupled for the two different directions, i.e. an inclination of the leg in one direction causes reaction moments in another direction and vice versa (see Fig. 20)

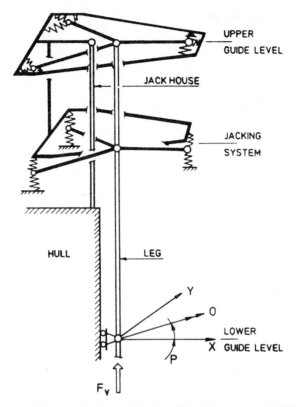

Fig. 16. A dedicated non-linear FEM model of the leg to hull interface.

In general the resulting spring characteristic is of a multi-linear nature.

This dedicated detailed model will be used to generate tables of rotational spring stiffness values for an appropriate range of vertical legloads, inclination orientation and inclination values. During the

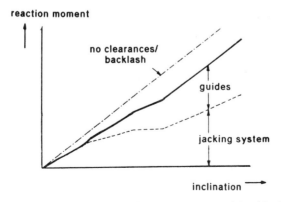

Fig. 17. Dependency on guide clearances and backlash.

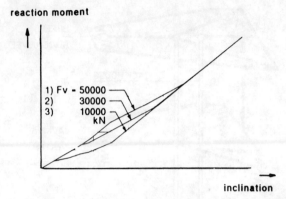

Fig. 18. Dependency on vertical leg reaction.

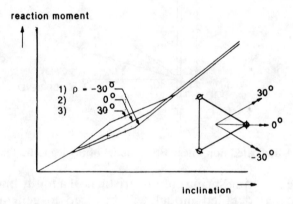

Fig. 19. Dependency on the inclination orientation.

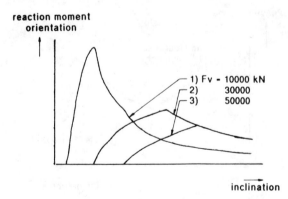

Fig. 20. Effect of coupling between directions.

simulation, for the momentary values of these variables, the stiffness is interpolated from these tables and used as the actual spring characteristic. The connection of each of the legs to the seabed is basically represented by five non-linear springs, two rotational and three translational.

The characteristics of this leg to seabed connection are very important for the overall behaviour of the jack-up. The question whether or not some fixity from the soil will be present has been the subject of several research projects and discussions. A consensus seems to be that when the foundation is vertically preloaded to a certain value, combinations of moments, vertical and horizontal forces can be accommodated under elastic conditions as long as the soil stress level does not exceed the stress level reached during preloading. For stress levels above this level the soil will yield and the characteristics, especially with respect to rotation, become plastic.[6] This means that the characteristics of the five springs at each footing are strongly interrelated.

To take into account the non-linear behaviour of the spudcan/soil interface the model as described in Ref. 14 has been implemented. The non-linear load displacement model takes into account the elastic deformation of the soil up to the bearing capacity. When the bearing capacity, calculated with well-known formulae (i.e. Brinck Hansen[15]), is exceeded, additional plastic deformation will occur. The ratio between horizontal, vertical and rotational deformation is determined by a flow rule and takes into account the ratio by which the bearing capacity is utilised by the horizontal, vertical and rotational (moment) footing reactions (see Fig. 21 taken from Ref. 14). Also the effect of a partially embedded spudcan can be taken into account, i.e. the increase in

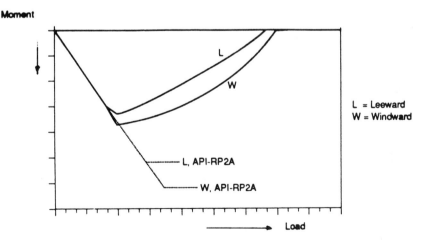

Fig. 21. Decrease in moment — fixity when approaching failure.

foundation area with increasing penetration. The model has been calibrated against results of model tests and FEM calculation.

Generally, three main sources of damping can be distinguished:

—structural damping;
—soil damping;
—hydrodynamic damping.

Damping properties are usually difficult to obtain. Little data on the magnitude of the damping are known although some full scale experiments have been performed.[16]

The common procedure for including damping properties in a time domain simulation is to apply Rayleigh-type damping. The damping matrix $[C]$ is then assembled from the stiffness matrix $[K]$ and mass matrix $[M]$ according to:

$$[C]: \alpha [M] + \beta [K]$$

with α and β following from typical damping ratios for the main modes.

The procedure described above has been implemented in the time domain program. Extra hydrodynamic damping can be included by calculating the fluid forces when applying the Morison equation using relative velocities/accelerations instead of absolute values. As soil damping, especially rotational, can be very important for the jack-up dynamic response (see for instance Ref. 17), additional rotational soil damping (dashpots) can also be included if required.

4.4 Statistics

The properties of interest to be established by the time domain analysis are Most Probable Maxima/Minima of the main parameters such as leg bending moments, footing reactions, etc., in a storm with a typical duration of 3 h. To limit the amount of computational effort, one would like to be able to estimate an extreme (for instance a 3 h extreme) from the simulated response time series of a simulation which is as short as possible. On the one hand the accuracy of the extreme estimates will require a maximum simulation length, on the other hand practical considerations ask for a limited simulation length.

If a linear relation between exitation and response exists, the nature of the response would be Gaussian, just as the input (wave) signal. The extremes of the response would then be distributed according to a Rayleigh distribution (assuming a narrow band process). However, due to the non-linearities in the system, for the loading (crest kinematics, the drag term in the Morison equation) as well as in the structural response (P-δ

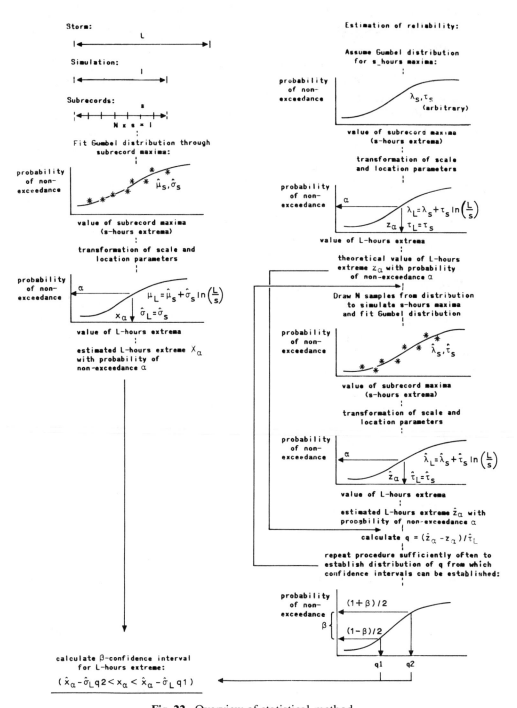

Fig. 22. Overview of statistical method.

effect, non-linear leg to hull and footing behaviour), the extremes of calculated responses will in general deviate from the Rayleigh distribution.[18]

The common procedure in case the exact description of the distribution functions of the extreme is unknown, is to assume a parameterised distribution function and to fit the parameters of the adopted distribution function to the simulated record, and use these parameter values to estimate extreme responses. The proper selection of the distribution is very important; the adopted distribution, especially, should be a good approximation to the real (unknown) distribution particularly in the tail of the distribution (extreme responses) although, unfortunately, few fitting points will be available in that part.

The method as developed for this project is described below (see also Fig. 22). To estimate the distribution of extremes of a variable in a certain period L(L-hours extremes) from a simulated time record of length l, the procedure is as follows.[19]

First the time record of length l is subdivided into a number of N subrecords with lengths $= l/N$. For each subrecord the maximum value is determined, giving N subrecord extremes which then are ranked in increasing order. With a length of each subrecord of approximately 20 times the important periods in the system, the subrecord extremes may be assumed to be independent.

In this way extreme value distributions may be considered to be limiting distributions of the subrecord extremes. Limiting distributions are always either Gumbel or Weibull distributions. Here the Gumbel distribution is selected, because it is relatively easy to handle:

$$F^s(x) = e^{-e^{\frac{-(x-\mu)}{\sigma}}} \tag{1}$$

where $F^s(x)$ being the probability that the s-hour maxima will exceed the value x and μ and σ being the location resp scale parameters.

The parameters of μ and σ can be estimated from the N subrecord maxima. Plausible estimates are the maximum likelihood estimates. The probability density of the observations as a function of μ and σ is maximised by setting the two derivatives of the likelihood with respect to μ and σ equal to zero.

By assuming the L-hour period to be built up from L/s independent s-hour intervals the distribution of the L-hours extreme can be estimated as:

$$F^L(x) = [F^s(x)]^{L/s} = e - \left(\frac{x - [\hat{\mu} + \hat{\sigma}\ln(L/s)]}{\hat{\sigma}}\right) \tag{2}$$

This is a Gumbel distribution with a location parameter $[\hat{\mu} + \hat{\sigma}\ln(L/s)]$ and a scale parameter $\hat{\sigma}$ with $\hat{\mu}$ and $\hat{\sigma}$ as the maximum likelihood estimates corresponding to the s-hour maxima. The value of the L-hour period extreme which will not be exceeded with a probability of α, called x_α can then be estimated as:

$$\hat{x}_\alpha = \mu: - \sigma: \ln\left[\ln(1/\alpha)^{s/L}\right] \tag{3}$$

This procedure results in an estimate for the L-hour extreme $\hat{x}_\alpha$.

However, these estimates themselves are subject to a statistical distribution, the statistical properties of which are of interest. That is, what is the range in which the L-hours $\hat{x}_\alpha$ extreme lies with a certain probability. So it is necessary to establish the confidence interval (q_1, q_2) according to:

$$\text{Probability } (q_1 < q < q_2) = \beta$$

where

$q = (\hat{z}_\alpha - z_\alpha)$;
$\hat{z}_\alpha = $ estimated α-quantile of the L-hours extreme;
$z_\alpha = $ theoretical (exact value).

Because it is difficult to determine the distribution of q analytically a numerical simulation procedure is followed. Instead of using the sub-record maxima from the time domain simulation, subrecord maxima are simply drawn from a Gumbel distribution with arbitrarily chosen location and scale parameters (λ and τ). Using the procedure described before, the L-hours extremes $\hat{z}_\alpha$ can be estimated from the subrecord maxima and can be compared to the theoretical value of z_α:

$$z_\alpha = \lambda - \tau \ln\left[\ln(1/\alpha)^{s/L}\right]$$

By repeating this procedure sufficiently often the distribution of $q = (\hat{z}_\alpha - z_\alpha)/\hat{\tau}$ can then be determined in a form independent of the location and scale parameters λ and τ.

Values of q_1 and q_2 can then be found and applied to the simulation results to find confidence intervals for each estimated L-hour extreme $\hat{x}_\alpha$ as:

$$\text{Confidence } (\hat{x}_\alpha - \hat{\sigma}q_2 < x_\alpha < \hat{x}_\alpha - \hat{\sigma}q_1) = \beta$$

With the above procedure developed and built into the program, the result will be presented as a Most Probable Maximum (MPM) with a corresponding confidence interval. In using the program for a range of jack-ups in a range of environmental conditions, experience can be built up with regard to the required ratio between the typical 3 h storm and the simulation period in the time domain analysis.

5 PROGRAM DESCRIPTION

The total program consists of a suite of four programs (Fig. 23). The communication between the programs is by external files. The functions of the various programs are described below:

5.1 Input program

The input program will interactively accept data from the user and generate a general datafile containing all needed structural data of the jack-up and the environmental data.

Much attention has been paid to enable the user to supply input at the level at which the data are available to him. For example, the stiffness properties of the beam elements representing the legs in the structural model are generally not directly available to the user. So, the program calculates leg inertia, etc. from the leg geometry and scantlings, which can be obtained from the leg drawings. In the same way, from the leg geometry and member properties the program calculates 'stick' hydro-dynamic properties.

The program reads the problem data file and transforms the structural data into an internal format for the solver. For the whole period to be simulated wave kinematics are generated and stored in the kinematics data file. During the actual time domain simulation the solver will fetch the wave kinematics for each time step from this file. The control datafile will contain data controlling the simulation execution, such as time step, integration parameters, etc.

User input is collected via several screen forms (Fig. 24). The program has been further optimised in user-friendliness by offering database facilities to prevent retyping, etc. The results of the files are written to a problem datafile.

5.2 Pre-processor

The main tasks of the pre-processor module are the generation of a structural datafile, a kinematical datafile and a control datafile.

5.3 Solver

The solver performs the time integration which is based on the Kok-γ method.[20] In this the equations of motions are solved by evaluating the momentum balance at each time step by means of the Galerkin Residual method. The Newmark-β method, which is often applied in time domain integration, can be derived from the Kok-γ method by selecting

$$\beta = 1/4 - 1/12\,\gamma^2$$

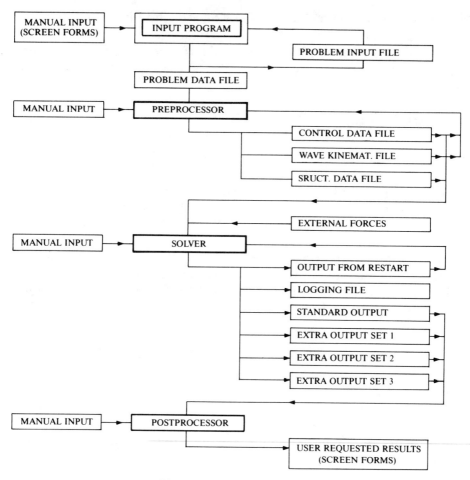

Fig. 23. Layout of program.

JUPTD M.S.C. b.v. V 1.00
Unit: default title 1
Case: default title 2

Command: ▲ ▲ ▲ ▲
Screen: 60

Definition of leg crosssection:
————————————

leg shape: 3   Options: 3 triangular leg
                         4 square leg

chords:

no:	chord type (pipe/composed)	orientation [degr.]
1	COMP	90.0
2	COMP	150.0
3	COMP	210.0

chord distance: 12.00 m   bracing type: X   Options: X for X-braces
bay height   :   5.97 m   K for K-braced

Fig. 24. Typical inputscreen.

By taking a value of γ equal to 0 the Kok-γ method is unconditionally numerically stable.

During the simulation, the main results will be written to the results file. Results are, for instance, total base shear, footing reactions, leg bending moments, etc.

5.4 Post-processor

Input for the post-processor are the results files generated by the solver containing time histories for the most relevant variables such as bending moments, leg to hull reactions, footing reactions, etc. From each time history the statistical properties can be determined and extreme values (most probable maxima) estimated according to the procedure described earlier.

The post-processor also enables the user to select parts of this file (for instance only one output variable for a limited simulation length) and write the results to other files, to conduct further processing by other programs outside the time domain package.

6 CONCLUDING REMARKS

The method described in this paper is presently being coded for use on 'Personal Engineering Workstation' type computers. It is expected that a 4–10 MIPS workstation will provide results within an acceptable time frame. At the time of the presentation of this paper, coding of the method is nearly completed, so results cannot, unfortunately, be shown as yet.

ACKNOWLEDGEMENTS

This development project has been made possible thanks to contributions both financially and technically of the Netherlands Foundation for Coordination Maritime Research (CMO) and Shell Internationale Petroleum Maatschappij (SIPM).

REFERENCES

1. Grenda, K. G., *Wave Dynamics of Jack-up Rigs,* OTC 5304, Houston, May 1986.
2. Pascoe, K. J. *et al., Self-Elevating Mobile Offshore Drilling Units, An Operators Viewpoint,* OTC 5359, Houston, April 1989.

3. Bradshaw, I. J., Jack-up structural behaviour and analysis methods. Paper presented at the International Conference on Mobile Offshore Structures, Shell International Petroleum, Maatshappij BV, The Netherlands.

4. Leyten, S. F. & Efthymiou, M., *A Philosophy for the Integrity Assessment of Jack-up Units*, SPE 19236 Offshore Europe, Aberdeen, 1989.

5. Gründlehner, G. J., The development of a simple model for the deformation behaviour of leg to hull connections of jack-ups. Thesis, TU Delft, August 1989.

6. Hattori, Y., Ushio, M., Ishihama, T. & Kawamura, T., Experimental study of foundation stability of jack-up rigs. *Conference on the Jack-up Drilling Platform*, City University, London, September 1985. Hitachi Zosen Corporation.

7. Sarpkaya, T. & Isaacson, M., *Mechanics of Wave Forces on Offshore Structures*, Van Nostrand Reinhold Co., New York, 1981.

8. Rodenbusch, G., *Random Directional Wave Forces on Template Offshore Structures*, OTC 5098, Houston, 1986.

9. Sing, S., Cash, R., Harris, D. & Boribond, L. A., Wave forces on circular cylinders with large excrescences at low Keulegan and Carpenter number. NMI report R133 (OT-R-8209), March 1982.

10. Tucker, M. J., Challenor, P. G. & Carter, D. J. T., Numerical simulation of a random sea: a common error and its effect upon wave group statistics. *Applied Ocean Research*, 6(2) (1984).

11. Mes, J., Paper waves for deep water platform design. *J. Mes., Petroleum Engineer*, May (1977).

12. Hudpeth, R. T. & Chen, M. C., Digital simulation of non-linear random waves. *Journal of the Waterway, Port, Coastal and Ocean Division*, Feb. (1979).

13. Rodenbusch, G. & Forristall, *An Empirical Model for Random Directional Wave Kinematics near the Free Surface*, OTC 5099, Houston, May 1986.

14. Schotman, G. J. M., *The Effects of Displacements on the Stability of Jack-up Spudcan Foundations*, OTC 6026, Houston, May 1989.

15. Brinck Hansen, J., A revised and extended formula for bearing capacity, Bulletin No. 28, The Danish Geotechnical Institute, Copenhagen, 1979.

16. Hattori, Y. et al., *Full-scale Measurement of Natural Frequency and Damping Ratio of Jack-Up Rigs and Some Theoretical Considerations*, OTC 4287, Houston, May 1982.

17. Hansen, H. T., Svenningsen, K. B. & Mathieu, P., Dynamics of jack-up platforms, Part report No. 4: time integration analyses. DNV Report 82–0797, Oslo, 1982.

18. Brouwers, J. J. H. & Verbeek, P. H. J., Expected fatigue damage and expected extreme response for Morison-type wave loading. *Applied Ocean Research*, 5(3) (1989).

19. Linssen, H. N. & Rienstra, S. W., Estimation of the distribution of extreme responses to random seas of offshore structures. Eindhoven Technical University Report IWDE 89–03, Eindhoven, June 1989.

20. Kok, A. W. M., Pulses in finite elements. *Proc. First Conference on Comp. in Civ. Eng.*, New York, 1981.

A Study of Jack-Up Leg Drag Coefficients

N. Pharr Smith & Carl A. Wendenburg

Marathon LeTourneau Marine Company, 600 Jefferson, Suite 1900, Houston, Texas 77002, USA

ABSTRACT

The accurate prediction of drag coefficients for truss-type jack-up legs is an important part of a unit's design. Over the past 14 years, a series of wind-tunnel tests have been conducted on both square and triangular cross-section leg configurations. A calculation method which closely fits the model test data has also been adopted. A previous paper described the early testing and the MMEC calculation method for square legs. Since then, testing and research have continued on both square and triangular leg designs. This paper describes the more recent testing performed on square legs, and it details the testing and the MMEC method for triangular legs. In addition, the effects of cornerpost design and surface roughness are discussed. Several example calculations are provided in Appendices.

Key words: jack-up, truss, leg, drag, coefficient, roughness.

1 INTRODUCTION

One of the many challenges facing designers of self-elevating mobile offshore drilling units continues to be the determination of accurate drag coefficients for truss-type legs. This aspect of jack-up design has received a great deal of attention in the past 10 years, and there has been much discussion about the best way to determine leg drag forces. The regulatory agencies provide guidelines for calculating drag coefficients, but these formulae have been shown to be somewhat inconsistent and difficult to interpret. While wind-tunnel testing has been used by some

185

designers, this approach also has drawbacks. Tunnels with the correct combination of size, wind speed, and balance capabilities are rare. The best tunnels are also very expensive. In addition, tests must be carefully planned to ensure that appropriate Reynolds numbers are reached. Even the best wind-tunnel tests cannot predict the effects of oscillatory flow on drag forces. These effects must be quantified in wave basins. Unfortunately, there are a limited number of wave basins large enough to test full leg models, and even these few basins cannot reach full-scale Reynolds or Keulegan–Carpenter numbers. In fact, wind tunnels can reach far higher Reynolds numbers than the best wave basins can.

A combination of extensive wind-tunnel testing and straightforward calculations has been used to approach this difficult problem. The testing has addressed both leg truss components and full-leg configurations. It has been performed in tunnels which provide measurements at the highest Reynolds numbers currently obtainable. The calculation method is a simple building block technique based on the test results, and it closely fits the data for a wide variety of leg types. While the effects of Keulegan–Carpenter number are not incorporated, there is evidence that the use of supercritical steady-flow drag coefficients is at least conservative.

In 1983, a paper was published describing both the wind-tunnel test program and the MMEC calculation method.[1] At that time, five major tests had been performed, concentrating primarily on square cross-section leg designs. Since then, six more tests have been undertaken. These tests had two common goals: to reach the maximum Reynolds numbers possible for several of the designs tested in the earlier program, and to investigate in detail the drag characteristics of triangular cross-section leg designs. As a result of this testing, the calculation method has been adapted to allow the prediction of drag coefficients for triangular legs. As a by-product of this testing, the significant effect that roughness has on the drag of tubular legs was noted. 'Roughness' in this case applies not only to marine growth but also to seemingly minor design and construction details.

The purpose of this paper is to update the information given in Ref. 1 and to present the results of some of the testing performed since that time. The test results are compared with the values obtained from the calculation method. In addition, the effects of roughness are described and discussed.

2 SQUARE LEGS

The previous paper[1] presented some of the results of testing at the Texas Engineering Experiment Station (TEES) at Texas A & M University. As explained in that paper, the TEES wind tunnel has a unique combination of size, wind speed, and balance capabilities, which allow testing at very high Reynolds numbers. It was possible to develop a sizable body of consistent data at TEES, with confidence that supercritical flow on the models had been realised. Nevertheless, ways to reach higher Reynolds numbers continued to be investigated.

Two square-leg designs were chosen for further testing: the Gorilla leg and the 116-Class leg.

2.1 The Gorilla leg

In the early testing program, the Gorilla-leg design was tested several times using a 1/12-scale model. In an effort to corroborate the 1/12-scale tests, a relatively large-scale half-plane model was employed. This approach allowed the use of a larger model in the same tunnel test section. To reduce the forces on the balance, a single bay was instrumented (Fig. 1). As shown in Fig. 2, this method enabled more than twice the 1/12-scale Reynolds numbers to be achieved. These tests supported the earlier conclusions: the 12-scale test results are conservative, and the MMEC calculation method seems to predict the supercritical C_D (Appendix A).

2.2 The 116-Class leg

The early testing program produced a great deal of data on the 116-Class leg. (In Ref. 1, the 116 leg was described as a square leg having 'triangular

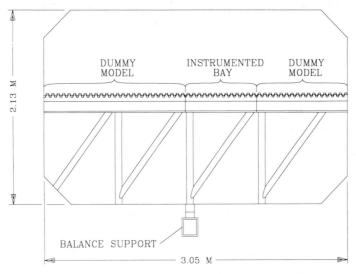

Fig. 1. Large-scale half-plane Gorilla-leg model in the TEES tunnel — streamwise view.

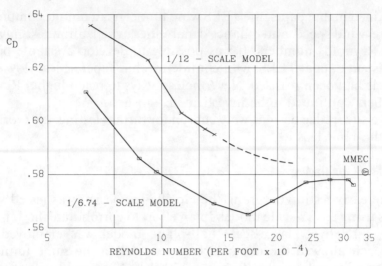

Fig. 2. TEES test results for Gorilla leg at two model scales compared with the MMEC calculation method results. (1 ft = 0·3048 m).

cornerposts with gussets.') Because the 116 leg was the basis for several parametric studies, it was used as the subject of a very high Reynolds-number test.

The site for the new testing was the National Aerospace Laboratory (NLR) in Amsterdam. Like the TEES facility, the NLR high-speed tunnel (HST) possesses a favorable combination of large test-section size (1·6 by 2·0 m), high wind speed (up to 1·27 Mach), and high balance limits (up to 33 kN). In addition, the NLR HST has the capability to vary the test-section pressure (from 0·125 to 4·0 atm), and the floor and ceiling of the test section are slotted to reduce wall interference and blockage effects. For these tests, the Mach number was varied from 0·20 to 0·45, and the pressure was varied from 1 to 4 atm. This program resulted in a Reynolds-number range of 8×10^6–35×10^6 per meter.

To maximize model size while keeping blockage to a minimum, a test set-up common in the aerospace industry was used: a sidewall model support with three-dimensional models (Fig. 3). Each leg model was tested first with three repeating sections (bays) installed in the tunnel; then a fourth bay was added, and the measurements were repeated. The difference between the two runs represented the drag characteristics of a single bay. By using this technique, both three-dimensional end effects and tunnel wall boundary layer effects were minimized. The run program consisted of Reynolds number, Mach number, and yaw-angle sweeps,

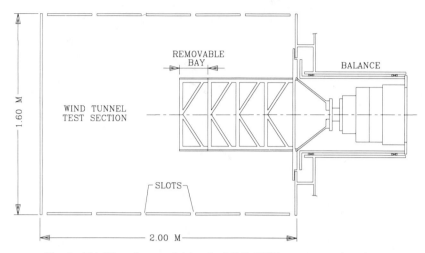

Fig. 3. 116-Class leg model in the NLR HST — streamwise view.

including several overlapping conditions to check for compressibility effects and repeatability.

A summary of the 116-Class leg testing is illustrated in Figs 4 and 5. The three 1/12-scale tests performed at TEES in 1975, 1980, and 1983 show very good repeatability (less than 2% variance). While the NLR test resulted in somewhat lower C_D values, there was an upward trend evident at the highest Reynolds numbers tested. This behavior is typical of trusses composed of cylindrical members. Since there was no attempt to stimulate supercritical flow in the NLR tests (e.g. by using trip strips or roughness on the cylinders), the results must be extrapolated to full-scale as shown in Fig. 5. This extrapolation indicates a full-scale (100-knot wind) C_D at zero degrees of about 0·610 as compared with 0·665 for the TEES tests and the MMEC method (Appendix B). Considering the average drag coefficient for the three legs as they are oriented on the jack-up, the NLR tests indicate a value of 0·697 while the MMEC method predicts a value of 0·745. Again, the 1/12-scale tests are shown to be conservative, and the MMEC method predicts a relatively accurate supercritical C_D (within about 6% of the NLR tests).

The early testing also investigated a 116-type leg with smooth cylindrical chords replacing the typical 116 triangular cornerposts. The tests on this configuration were similar to the Gorilla-leg tests; both a 1/12-scale model and a large-scale half-plane model were used. The results of these tests are shown in Fig. 6. The wind-tunnel tests indicate a C_D at zero degrees of 0·471, and the MMEC method predicts 0·465 (Appendix C).

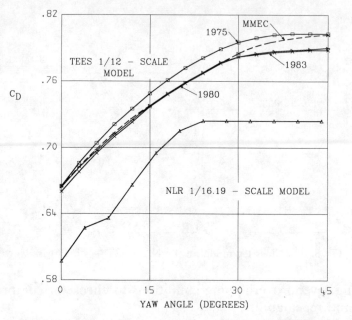

Fig. 4. TEES and NLR test results for 116-Class leg compared with the MMEC calculation method results.

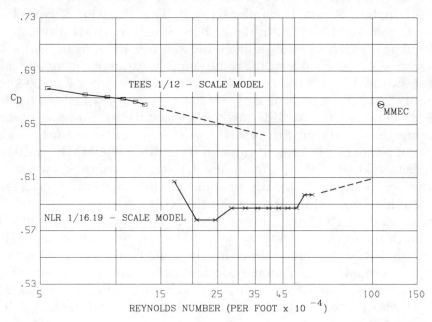

Fig. 5. TEES and NLR test results for 116-Class leg compared with the MMEC calculation method results. (1 ft = 0·3048 m).

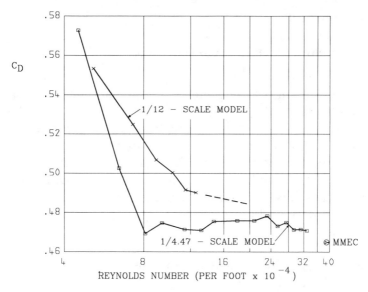

Fig. 6. TEES test results for 116-Type leg with cylindrical chords compared with the MMEC calculation method results. (1 ft = 0·3048 m).

3 TRIANGULAR LEGS

Testing of triangular-leg designs began in 1979. When the previous paper was prepared, however, the triangular-leg data had not been sufficiently analyzed to include these results. Since that time, several more tests have been performed both at TEES and at the NLR HST, and these tests have provided the opportunity to calibrate the MMEC method for triangular legs.

The general form of the MMEC method for triangular legs is

$$C_D = K_{\beta\Delta}[\Sigma_w A_i C_{di} + \eta_\Delta \Sigma_l A_j C_{dj}]/LW \qquad (1)$$

where

$K_{\beta\Delta}$	=	orientation factor = 1·0
β	=	yaw angle
w	=	windward
l	=	leeward
A_i	=	component projected areas of windward face
A_j	=	component projected areas of leeward faces
C_{di}	=	drag coefficients for components of windward face
C_{dj}	=	drag coefficients for components of leeward face
C_{dwc}	=	drag coefficient for windward cornerposts
C_{dlc}	=	drag coefficient for leeward cornerpost

N. Pharr Smith, Carl A. Wendenburg

L = length of bay
W = width of bay
η_Δ = shielding factor = $0 \cdot 8 \, (1 \cdot 1 - C_{dwc} \, \phi)$
ϕ = $\Sigma_w \, A_i / LW$

(A zero degree yaw angle corresponds to an orientation with the windward face perpendicular to the wind.)

As described in Ref. 1, this formulation is not a new approach. It borrows ideas and values from the classification society codes and several important papers.[2-8] The MMEC method is intended to be a calculation method specifically for jack-up legs with coefficients picked to fit the test data. It is simply a refinement of the more general methods presented by others, and it is offered as a way to reduce the need to test every leg design.

3.1 Triangular legs with triangular cornerposts

The first tests of triangular legs were performed at TEES on 1/12-scale models of the 150–44 and 82-Class leg designs. Figure 7 shows the Reynolds number data from these tests. At a yaw angle of zero degrees, the MMEC method appeared to predict reasonable C_D values for these legs (Appendices D and E). Figure 8 shows the yaw-angle sweep data for

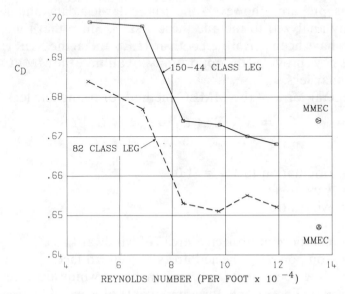

Fig. 7. TEES test results for two triangular-leg designs with triangular chords (at zero degrees yaw angle) compared with the MMEC calculation method results. (1 ft = $0 \cdot 3048$ m).

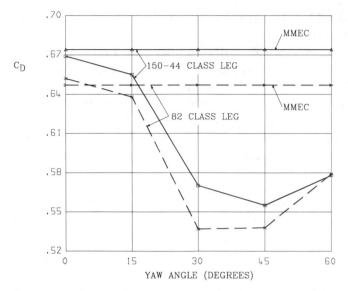

Fig. 8. TEES test results for two triangular-leg designs with triangular chords compared with the MMEC calculation method results.

the same two leg designs. As can be seen, the yaw-angle behavior for triangular legs is not as simple as that for square legs. This behavior becomes even more complex when leg solidity and cornerpost type are varied. For this reason, previous calculation methods have attempted to predict a single conservative value for all yaw angles. This approach is chosen here, setting $K_{\beta\Delta}$ equal to 1·0. Since triangular legs are usually all oriented in the same direction on the jack-up, this design approach is not excessively conservative. Of course, for site-specific applications, full advantage could be taken by using the wind-tunnel data for the yaw angle in question.

3.2 Triangular legs with tubular cornerposts

The early tests on triangular legs with tubular chords were performed at TEES using 1/8·47-scale models. Two configurations were tested: a triangular leg with purely cylindrical chords, and the same leg with opposed racks added to the chords. In order to stimulate supercritical flow around the leg members, a lathe was used to knurl the surface of each cylinder. Knurling pattern and depth were chosen so as to trip the flow without unrealistically raising the drag coefficient, but preliminary tests on single, knurled cylinders showed that the member C_d value would be approximately 0·82. Since the correct full-scale C_d value is 0·70, the measured leg drag coefficient should be reduced to account for the added roughness.

Figure 9 shows the Reynolds-number plots for the two leg models, and Fig. 10 shows their yaw-angle behavior. For these two leg configurations, the MMEC method predicts an average value somewhat lower than the maximum value for each leg: about 5% lower for the leg with racks, and about 2% lower for the leg without racks (Appendices F and G).

In the quest for ever higher Reynolds numbers, a triangular-leg design was tested at the NLR HST. This time a more popular chord design was chosen: instead of simply adding racks to a cylinder, a leg with more realistic split-tube opposed-rack cornerposts was tested (see Fig. 11). For this test, the leg components were left smooth. Although the NLR HST allowed relatively high Reynolds numbers to be reached, the results must still be extrapolated to full scale. Figures 12 and 13 show the results of these tests. In this case, the MMEC method predicts an average C_D value approximately 5% above the test results (Appendix H).

In summary, the testing program for triangular legs has explored a fairly broad range of leg solidities and cornerpost types. Also, the MMEC method has proved to predict reasonable average C_D values for these legs (within $\pm 5\%$). These tests, however, raised two areas of uncertainty: the effect of minor changes in tubular cornerpost geometry, and the effect of roughness.

4 TUBULAR CORNERPOSTS

While most of the testing concentrated on full-leg models, tests have also been performed on several isolated cornerpost models. Testing isolated cornerposts allowed much higher Reynolds numbers to be reached and provided the required input for the MMEC calculation method. These tests also provided insight into the drag effects of relatively minor chord-design changes without having to build full-leg models.

When testing on tubular cornerposts first began at TEES, unexpected differences between the test results and the C_d values recommended by the regulators were realized. As more tests were performed and the data analyzed more carefully, however, two conclusions were reached. First, tubular cornerposts must be tested at full-scale Reynolds numbers. These chord designs are especially sensitive when the flow is parallel to the rack. Second, the cornerpost shape and tooth profile are much more important than first realized. In this regard, idealized shapes must be viewed with some skepticism. Particularly surprising was the great influence of the dimension from the root of the teeth to the chord

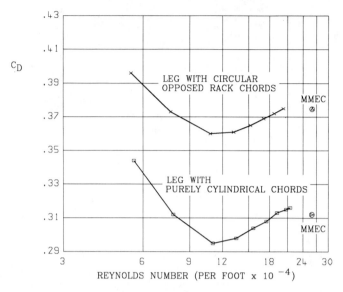

Fig. 9. TEES test results for triangular leg with two chord designs (at zero degrees yaw angle) compared with MMEC calculation method results. (1 ft = 0·3048 m).

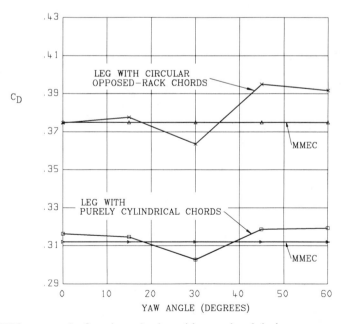

Fig. 10. TEES test results for triangular leg with two chord designs compared with the MMEC calculation method results.

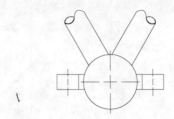

CIRCULAR CHORD WITH OPPOSED RACKS ADDED

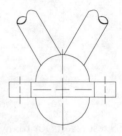

Fig. 11. Two basic tubular cornerpost
types.

SPLIT–TUBE OPPOSED–RACK CHORD

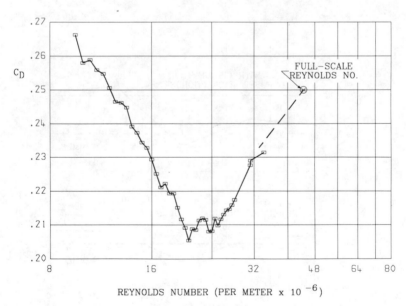

REYNOLDS NUMBER (PER METER x 10^{-6})

Fig. 12. NLR test results for triangular leg with split-tube opposed-rack chords (at 30°
yaw angle).

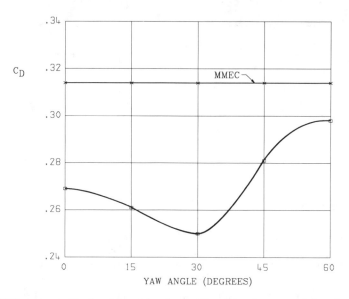

Fig. 13. NLR test results for triangular leg with split-tube opposed-rack chords at full-scale Reynolds number compared with the MMEC calculation method results.

diameter. Figures 14 and 15 give a summary of three of the tested shapes compared with a smooth cylinder.

5 SURFACE ROUGHNESS

The effects of surface roughness on the drag coefficient of an isolated cylinder are well known. What is not so well understood is the effect roughness has on the C_D for truss-type legs. Unfortunately, very little testing has been performed on roughened legs. The analyst is left, therefore, with the job of modifying the available building-block calculation methods to account for marine growth. While most analysts agree that both increased projected areas and increased member C_d values should be used in the calculations, some designers feel that an allowance for increased area is all that is required.

Two leg designs with surface roughness were tested: the standard 116 leg, and a 116-type leg with purely cylindrical chords. The results of these tests have been compared with calculated values using the MMEC method (Appendices B and C) and are shown in Table 1.

Several conclusions can be drawn from this table. First, roughness has a greater effect on the all-cylindrical leg than on the leg with triangular chords. The tests indicated a 32% increase in drag for the all-cylindrical

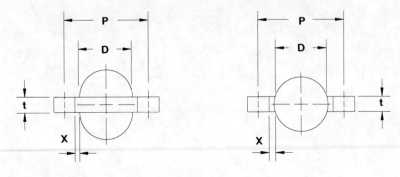

Cornerpost model	P	D	t	x
Split-tube opposed-rack No. 1	28·50	18·00	5·00	1·63
Split-tube opposed-rack No. 2	22·56	15·00	5·00	0·37
Circular with opposed racks	28·50	18·00	5·00	1·63

Fig. 14. Three tubular cornerpost designs tested at TEES.

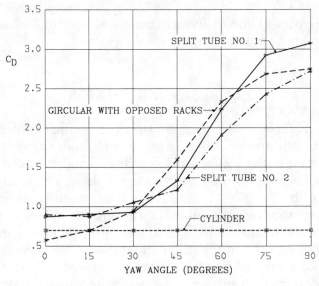

Fig. 15. TEES test results for three tubular cornerpost designs compared with a circular cylinder. Reference length = D (see Fig. 14).

leg, and only a 10% increase for the standard 116 leg with triangular chords. Second, the MMEC method predicts the increase in drag fairly accurately when both increased projected area and increased C_d for the

TABLE 1
Effect of Surface Roughness on Drag Coefficients for Two Leg Designs

Leg design	Roughness	Test result	MMEC method result	
			Area & C_d	Area only
116-Class leg	Smooth	0·665	0·665	—
116-Class leg	Rough	0·730	0·737	0·666
All-cylindrical leg	Smooth	0·471	0·465	—
All-cylindrical leg	Rough	0·620	0·643	0·480

cylinders are used: the difference between the test results and the calculated values is less than 4%. (Note that for the roughness used in these tests, the appropriate C_d value for the roughened cylinders is 1·0.) Third, when only projected area is increased in the calculations, the C_D value is underpredicted, by 23% for the all-cylindrical leg and by 9% for the standard 116 leg. While it is true that wave theories are generally thought to predict conservative particle velocities, it is clear that this safety margin may not cover the large percentage increase in drag caused by marine growth on tubular legs. In fact, some tubular triangular leg designs that have been explored appear to suffer an increase in C_D of greater than 50% when roughness is added. This is an area which definitely deserves more research.

6 CONCLUSION

This study has investigated the drag coefficients for a broad variety of truss-type jackup legs. Both square and triangular cross-section legs have been tested and analyzed. The leg designs have ranged from idealized all-cylindrical models to actual configurations with years of operating experience. The MMEC method proposed in 1983[1] has shown good agreement with the test results for both square and triangular legs. This calculation method has been proposed as an alternative to testing, especially when wind-tunnel test results for similar designs are available. The testing program has indicated that seemingly minor design details on tubular chords can have a significant effect on the leg drag coefficient. For this reason, idealized cornerpost C_d values should be critically evaluated. The testing has also shown that surface roughness can substantially increase the leg drag coefficient, especially for tubular leg designs. Moreover, when evaluating the effect of marine growth, the calculations have indicated the importance of considering not only the increased projected area but also the increased member drag coefficients. Needless to say, more research is needed in the areas of marine growth and Keulegan–Carpenter number effects.

REFERENCES

1. Smith, N. P., Lorenz, D. B., Wendenburg, C. A. & Laird II, J. S., A study of drag coefficients for truss legs on self-elevating mobile offshore drilling units. *Transactions, The Society of Naval Architects and Marine Engineers*, **91** (1983).
2. Roshko, A., Experiments on the flow past a circular cylinder at very high Reynolds number. *Journal of Fluid Mechanics*, **10** (1961).
3. American Bureau of Shipping, *Rules for Building and Classing Mobile Offshore Drilling Units (1980)*. American Bureau of Shipping, New York, 1980.
4. Det norske Veritas, *Rules for Classification of Mobile Offshore Drilling Units*. Det norske Veritas, Oslo, Norway, 1982.
5. British Standard's Institution, *Draft Code of Practice — Lattice Towers — Loading*. B.S.I., London, 1978.
6. Engineering Sciences Data Unit, *Item Number 75011: Fluid Force on Lattice Structures*. Engineering Sciences Data Unit, London, 1975.
7. Whitbread, R. E., The influence of shielding on the wind forces experienced by arrays of lattice frames. In *Proceedings of the 5th International Conference on Wind Engineering*,
8. Hoerner, S. F., *Fluid-Dynamic Drag*. Brick Town, New Jersey, 1965.

APPENDIX A: GORILLA LEG

$$C_{dwc} = 1 \cdot 882 \qquad C_{dlc} = 1 \cdot 573$$
$$\phi = (102 \cdot 81 + 15 \cdot 0 + 114 \cdot 5)/(16 \cdot 776) \ (46) = 0 \cdot 301$$
$$\eta = (1 \cdot 1) - (1 \cdot 882) \ (0 \cdot 301) = 0 \cdot 5334$$
$$C_D = \{[(102 \cdot 81) \ (1 \cdot 882) + (15) \ (2) + (114 \cdot 5) \ (0 \cdot 7)]$$
$$+ (0 \cdot 5334) \ [(102 \cdot 81) \ (1 \cdot 573) + (15) \ (2) + (114 \cdot 5) \ (0 \cdot 7)]\}$$
$$/(16 \cdot 776) \ (46)$$
$$= 0 \cdot 581 \text{ (zero degrees yaw angle)}$$

APPENDIX B1: 116-CLASS LEG (NO ROUGHNESS)

$$C_{dwc} = 2 \cdot 012 \qquad C_{dlc} = 1 \cdot 625$$
$$\phi = (57 \cdot 92 + 6 \cdot 0 + 56 \cdot 65)/(11 \cdot 182) \ (30) = 0 \cdot 3594$$
$$\eta = (1 \cdot 1) - (2 \cdot 012) \ (0 \cdot 3594) = 0 \cdot 3769$$
$$C_D = K_\beta \{[(57 \cdot 92) \ (2 \cdot 012) + (6) \ (2) + (56 \cdot 65) \ (0 \cdot 7)]$$
$$+ (0 \cdot 3769) \ [(57 \cdot 92) \ (1 \cdot 625) + (6) \ (2) + (56 \cdot 65) \ (0 \cdot 7)]\}$$
$$/(11 \cdot 182) \ (30)$$

$$
\begin{aligned}
&\qquad\quad = \; K_\beta \,(0{\cdot}665)\\
K_\beta &\quad= \; 1 + (0{\cdot}3594)\,(0{\cdot}5728)\,(\sin 2\beta)^{0{\cdot}9}\\
C_D\,(0) &\quad= \; 0{\cdot}665\\
C_D\,(15) &\quad= \; 0{\cdot}738\\
C_D\,(30) &\quad= \; 0{\cdot}785\\
C_D\,(45) &\quad= \; 0{\cdot}802
\end{aligned}
$$

Average for three legs

$$
\begin{aligned}
&= \; (0{\cdot}665 + 0{\cdot}785 + 0{\cdot}785)/3\\
&= \; 0{\cdot}745
\end{aligned}
$$

APPENDIX B2: 116-CLASS LEG (WITH ROUGHNESS)

Increased area and C_d

$$
\begin{aligned}
C_{dwc} &= 2{\cdot}012 \qquad C_{dlc} = 1{\cdot}625 \qquad C_{d(cylinder)} = 1{\cdot}0 \text{ (as tested)}\\
\phi &= (57{\cdot}92 + 6{\cdot}0 + 58{\cdot}78)/(11{\cdot}182)\,(30) = 0{\cdot}3658\\
\eta &= (1{\cdot}1) - (2{\cdot}012)\,(0{\cdot}3658) = 0{\cdot}3640\\
C_D &= \{[(57{\cdot}92)\,(2{\cdot}012) + (6)\,(2) + (58{\cdot}78)\,(1{\cdot}0)]\\
&\qquad + (0{\cdot}3640)\,[(57{\cdot}92)\,(1{\cdot}625) + (6)\,(2) + (58{\cdot}78)\,(1{\cdot}0)]\}\\
&\qquad /(11{\cdot}182)\,(30)\\
&= 0{\cdot}737
\end{aligned}
$$

Increased area only

$$
\begin{aligned}
C_{dwc} &= 2{\cdot}012 \qquad C_{dlc} = 1{\cdot}625 \qquad C_{d(cylinder)} = 0{\cdot}7\\
\phi &= 0{\cdot}3658\\
\eta &= 0{\cdot}3640\\
C_D &= \{[(57{\cdot}92)\,(2{\cdot}012) + (6)\,(2) + (58{\cdot}78)\,(0{\cdot}7)]\\
&\qquad + (0{\cdot}3640)\,[(57{\cdot}92)\,(1{\cdot}625) + (6)\,(2) + (58{\cdot}78)\,(0{\cdot}7)]\}\\
&\qquad /(11{\cdot}182)\,(30)\\
&= 0{\cdot}666
\end{aligned}
$$

APPENDIX C1: 116-TYPE ALL-CYLINDRICAL LEG (NO ROUGHNESS)

$$
\begin{aligned}
C_{dwc} &= C_{dlc} = 0{\cdot}7\\
\phi &= (120{\cdot}95)/(11{\cdot}182)\,(29{\cdot}44) = 0{\cdot}3674\\
\eta &= (1{\cdot}1) - (0{\cdot}7)\,(0{\cdot}3674) = 0{\cdot}8428\\
C_D &= [(120{\cdot}95)\,(0{\cdot}7) + (0{\cdot}8428)\,(120{\cdot}95)\,(0{\cdot}7)]/(11{\cdot}182)\,(30)\\
&= 0{\cdot}465 \text{ (zero degrees yaw angle)}
\end{aligned}
$$

APPENDIX C2: 116-TYPE ALL-CYLINDRICAL LEG
(WITH ROUGHNESS)

Increased area and C_d

$$
\begin{aligned}
C_{dwc} &= C_{dlc} = C_{d(cylinder)} = 1\cdot0 \text{ (as tested)} \\
\phi &= (125\cdot50)/(11\cdot182)\,(29\cdot44) = 0\cdot3812 \\
\eta &= (1\cdot1) - (1\cdot0)\,(0\cdot3812) = 0\cdot7188 \\
C_D &= [(125\cdot50)\,(1\cdot0) + (0\cdot7188)\,(125\cdot50)\,(1\cdot0)]/(11\cdot182)\,(30) \\
&= 0\cdot643
\end{aligned}
$$

Increased area only

$$
\begin{aligned}
C_{dwc} &= C_{dlc} = C_{d(cylinder)} = 0\cdot7 \\
\phi &= 0\cdot3812 \\
\eta &= (1\cdot1) - (0\cdot7)\,(0\cdot3812) = 0\cdot8332 \\
C_D &= [(125\cdot50)\,(0\cdot7) + (0\cdot8332)\,(125\cdot50)\,(0\cdot7)]/(11\cdot182)\,(30) \\
&= 0\cdot480
\end{aligned}
$$

APPENDIX D: 150–44-CLASS LEG

$$
\begin{aligned}
C_{dwc} &= 1\cdot97 \qquad C_{dlc} = 2\cdot06 \\
\phi &= (40\cdot95 + 61\cdot46 + 3\cdot65)/(11\cdot182)\,(24) = 0\cdot395 \\
\eta_\Delta &= (0\cdot8)\,[(1\cdot1) - (1\cdot97)\,(0\cdot395)] = 0\cdot257 \\
C_D &= \{[(0\cdot7)\,(40\cdot95) + (1\cdot97)\,(61\cdot46) + (2)\,(3\cdot65)] \\
&\quad + (0\cdot257)\,[(0\cdot7)\,(45\cdot22) + (2\cdot06)\,(26\cdot09) + (2)\,(3\cdot65)]\} \\
&\quad /(11\cdot182)\,(24) \\
&= 0\cdot674
\end{aligned}
$$

APPENDIX E: 82-CLASS LEG

$$
\begin{aligned}
C_{dwc} &= 1\cdot97 \qquad C_{dlc} = 2\cdot06 \\
\phi &= (36\cdot90 + 61\cdot46 + 7\cdot61)/(11\cdot182)\,(27) = 0\cdot351 \\
\eta_\Delta &= (0\cdot8)\,[(1\cdot1) - (1\cdot97)\,(0\cdot351)] = 0\cdot3268 \\
C_D &= \{[(0\cdot7)\,(36\cdot90) + (1\cdot97)\,(61\cdot46) + (2)\,(7\cdot61)] \\
&\quad + (0\cdot3268)\,[(0\cdot7)\,(48\cdot00) + (2\cdot06)\,(26\cdot09) + (2)\,(7\cdot61)]\} \\
&\quad /(11\cdot182)\,(27) \\
&= 0\cdot647
\end{aligned}
$$

APPENDIX F: TRIANGULAR ALL-CYLINDRICAL LEG

C_{dwc} = C_{dlc} = 0·82 (with roughness — as tested)
ϕ = (185·16)/(15·84) (54·19) = 0·216
η_Δ = (0·8) [(1·1) − (0·82) (0·216)] = 0·7383
C_D = [(0·82) (185·16) + (0·7383) (0·82) (173·77)]
/(15·84) (52·063)
= 0·312

APPENDIX G: TRIANGULAR LEG WITH CIRCULAR OPPOSED-RACK CHORDS

C_{dwc} = 0·9472 $\qquad$ C_{dlc} = 1·7348
$C_{d(cylinder)}$ = 0·82 (with roughness — as tested)
ϕ = (67·32 + 117·84)/(15·84) (54·19) = 0·216
η_Δ = (0·8) [(1·1) − (0·9472) (0·216)] = 0·7163
C_D = [(0·82) (117·84) + (0·9472) (67·32)]
+ (0·7163) [(0·82) (140·11) + (1·7348) (53·31)]
/(15·84) (52·063)
= 0·375

APPENDIX H: TRIANGULAR LEG WITH SPLIT-TUBE OPPOSED-RACK CHORDS

C_{dwc} = 0·8183 $\qquad$ C_{dlc} = 1·3888
ϕ = (49·80 + 38·64)/(12) (34·11) = 0·216
η_Δ = (0·8) [(1·1) − (0·8183) (0·216)] = 0·7385
C_D = [(0·7) (49·80) + (0·8183) (38·64)]
+ (0·7385) [(0·7) (61·79) + (1·3888) (29·38)]
/(12) (34·11)
= 0·314

Some Aspects of the Structural Design of a Mat Foundation for Various Soil Conditions

David B. Lorenz

Bethlehem Steel Corporation, Baltimore Marine Division, Sparrows Point Yard,
Sparrows Point, Maryland 21219, USA

ABSTRACT

There are several important structural design parameters in the analysis of a mat type structure for a jackup mobile offshore drilling unit. These consist of dimensions for the mat, soil foundation types, and structural loads. Methods for determining the proper dimension of the mat structure, modeling the structure and its loading cases, modeling the soil conditions and a method for identifying critical load cases for each element and load case is presented. By using a database approach for the finite element output for all of the various loading cases, a rational approach is presented which solves the problem of sorting and analyzing all of the elements and load cases, ensuring that all are included and none of the critical cases are missed.

Key words: structural analysis, foundation, jackup, finite element, mat structure, stress analysis.

1 INTRODUCTION

The range of soil conditions in which a jackup drilling rig can operate is becoming more important all of the time. As jackups are being designed for deeper water and harsher environments, and thus becoming more expensive, it is imperative that these designs have the capability to operate in widely varying soil conditions, water depths and environments. They should be 'world class' rigs, that are able to operate on any continental shelf in the world. Of course, this is not entirely possible due to some areas having ice, mud slides or other environmental features that are beyond the capabilities of present day jackup technology. However, a modern jackup should be designed to be able to operate in the greatest amount of continental shelf area possible.

With regard to soils, these jackups should have the capability to work in the softest clays and silts of the Gulf of Mexico and other river deltas, the relatively hard sands of the Arabian Gulf and other areas, the stiff clays throughout the world, and finally the sand waves of the North Sea and Cook Inlet.

A mat rig or a gravity foundation jackup provides the ability to work in a very wide range of soils due to its low bottom bearing pressure. This relatively low pressure is simple due to the fact that the mat covers a large area.

This paper will examine the need for wide range foundation capability. It will also present operational and structural criteria used in the design. It will study the operational and structural tradeoffs in mat design. Finally the techniques used in the structural design of a mat will be shown.

2 SOIL INFORMATION

The question of need for a jackup with the capability to operate on a wide range of soil conditions should be examined carefully. If a variety of bottom conditions did not exist to a sufficient extent, then there would be no need to design a rig for various soil conditions. For example, if there were only small amounts of 'special' soil conditions in the world, then a few purpose built jackups could be constructed to cover them. If, on the other hand, there were a great variety of soil conditions spread equally all over the continental shelves of the world, then it is obvious that a general purpose jackup with extremely wide soil capabilities would be desired.

In order to evaluate the need for this flexible jackup, an examination will be made of the bottom characteristics of two of the most demanding drilling areas of the world, the North Sea and the Gulf of Mexico. These areas have also been selected since there is a great deal of data available. In order for this not to be a treatise on soil properties, the paper will only deal with soil capabilities in general terms.

2.1 The North Sea

In general, the North Sea bottom is composed primarily of a veneer of sand on relatively hard clays. Data from several developed locations indicate that the undrained strength of these soils is generally in excess of 7 MPa (1 psf = 4.79×10^{-5} MPa). The overall slope of the bottom in water depths of up to 120 m is much less than $1°$.[2] There are sand waves located in some areas.

2.2 The Gulf of Mexico

Where the North Sea has fairly consistent bottom conditions, the Gulf of Mexico has a wide range. The type of soil can range from very soft to very hard clays, consolidated and loose silts and sands, to almost any combinations of these soils. Perhaps a better way to examine this is not to define the distribution of soils but rather the distribution of soil strengths. Figure 1 shows that the distribution of soil strength in the Gulf of Mexico is quite different from that of the North Sea. This table shows that approximately 30% of the area between 12 and 120 m has an undrained shear strength of less than 7·75 kPa and another 5% less than 1·92 kPa.[2]

The slope of the bottom is less than $1°$ for more than 99·8% of the area and generally much less than $1°$. There are problems in areas where mud slides can occur in some of the very soft soils, and where the soils have been deposited in layers of greatly differing soil strengths.

The information on the North Sea demonstrates no real advantage in having a foundation that is good for all soils. On the other hand, the information from the Gulf of Mexico shows that a foundation that can handle a wide range of soil types and strengths would have an advantage over one that could not.

The information shows that there is no overall problem with bottom

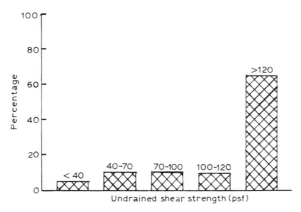

Fig. 1. Gulf of Mexico soil strength distribution (1 psf = $4·79 \times 10^{-5}$ MPa).

slope. Although there may be isolated instances where a slope of greater than 1° could exist. If a design is to be good for the North Sea, then a condition involving sand waves must be evaluated in the structural design. If the design is to be truly world class it must meet all of the criteria for soft soils, hard soils, and sand waves.

3 ENVIRONMENTAL CRITERIA

In order to design a mat type foundation, the loads that are applied to it must first be determined. For the purposes of this paper, the environmental loads for the storm condition are listed in Table 1.

These are typical North Sea criteria for 120 m water depth. Since the design areas of concern include the Gulf of Mexico, equivalent design storm criteria can be derived that will give approximately the same overturning moment and the same lateral loads on the structure. Table 2 shows this storm.

The environmental loads for the mat consist of overturning moments and lateral force due to the wave and current, and the same force components due to the wind.

TABLE 1
North Sea Design Storm

Water depth	120 m
Wave height	31·3 m
Wave period	18·5 s
Current	
Surface	0·77 m/s
Bottom	0·26 m/s
Wind speed	45 m/s

TABLE 2
Equivalent Gulf of Mexico Storm

Water depth	134 m
Wave height	20 m
Wave period	14·5 s
Current	
Surface	1·0 m/s
Bottom	0·26 m/s
Wind speed	65 m/s

4 STORM LOAD CALCULATIONS

Although this paper will only describe one set of calculations for a design storm, the same techniques are used to determine the forces and loads for the drilling condition, the going on location condition, the floating on the mat condition and all other storm conditions.

The forces and moments due to the wave and current are determined first using a Stokes 5th order wave program on an equivalent single pile. These initial forces and moments are used to determine the initial scantlings of the tower.

Secondly, a finite element model of the entire truss tower is generated and this model is used to determine the wave-tower phase angle to ensure maximum overturning moments, verify the drag coefficient used for the tower, verify the loads due to gravity are correct and included, and finally to verify that the forces and moments are equivalent to the equivalent single pile results. An example of this model is shown in Fig. 2. The finite element program used is ANSYS, which has wave loading capabilities and supports a Stokes 5th order wave. The current velocity is added to the wave particle velocity in this program.

The North Sea case shown in Fig. 3 demonstrates that almost 62·3 MN of shear are taken by the bulkheads in the mat. Since the horizontal forces are taken out as a couple between the deck and bottom plating of the mat, a deeper mat also allows these horizontal forces to produce less stress because of a larger couple. Several factors also dictate a large mat.

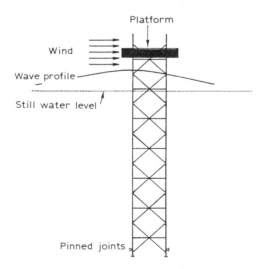

Fig. 2. Tower structural model.

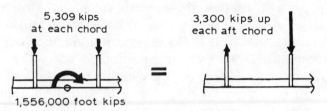

Fig. 3. Moment resolution of tower forces (1000 kips = 4·448 MN; 1000 ft kips = 1·356 MNm).

By keeping the mat as large as possible, it will have a lower bottom bearing pressure, and be more resistant to overturning.

The wind loads are calculated for wind from the bow and from the stern. The worser of these is chosen. In addition, the wind forces from the side and at 45° are also calculated. These loads are resolved about the four chords of the tower and are applied horizontally to the tower at the nodes of the platform. The platform is modeled as a mass element only and coupled to nodes on the tower. This is done in order to provide accurate weights to the model.

Finally loads for a 1° angle of tilt are applied vertically to the chords of the tower at the top nodes. These loads include the weight shift for the tower and the platform that would be caused by a 1° out of level angle.

The program is run for all storm load cases using a linear or elastic solution and an equivalent run is made using a non-linear large deflection model. This accounts for the *P* effect. The large deflection runs are updated with each iteration to account for the weight shift caused by large deflections of the tower. However, it should be noted that these deflections were on the order of 0·6 m or less and provided no major contribution to the overturning moment. A summary of these results in the form of overturning moments is provided in Table 3. Note that although the moments due to individual components, particularly wave and current and wind vary greatly, the total moment is approximately the same. The loads are then determined in a similar manner for all wind and wave directions.

Figure 3 demonstrates how these overturning moments would be resolved into the axial load components at the chord. These forces are then transmitted into the mat. In a similar manner, the moments about the chord and the horizontal forces are calculated. Note that although the overturning moment is negative in the resolved case, the weight and

TABLE 3
Summary of Storm Overturning Moments. Wind and Wave from
the Side

	Storm	
	North Sea	*Gulf of Mexico*
Wave and current	1 162 MNm[a]	530 MNm
Wind	601 MNm	1 214 MNm
1° out of level	279 MNm	283 MNm
P-δ	68 MNm	68 MNm
Total	2 110 MNm	2 095 MNm

[a] 1000 ft kips = 1·356 MNm.

spread of the mat are not yet included. These overturning moments and the lateral and vertical forces are resolved about the top of the first bay of the tower above the mat, so that they can be input into the mat structural model when it is run.

5 DESIGN TRADEOFFS

Structurally, it would be best to have a mat that is as deep as possible. This would allow a greater shear area to take out the large axial chord loads that are due to the large overturning moment. Ideally, the mat should cover as large an area as possible and be as deep as possible. Unfortunately, there are several other conditions that make the ideal solution impractical. First a large mat would require an extremely long cantilever to reach from the end of the platform to a drilling position beyond the edge of the mat. It would be impossible to do workover on an existing platform. The jackup could simply not get close enough to the base. Therefore a minimum footprint mat should be used.

A deep mat, although structurally ideal, does require more steel in construction. Additionally, a deeper mat is more susceptible to scour around its edges as the big bluff body will worsen any existing scouring problem. Finally, a deep mat can significantly increase the sliding force applied to the mat. In addition to the horizontal forces of the environment acting on the tower and the platform, there is a hydro-dynamic force acting on the mat which is directly proportional to the height of the mat structure. In Fig. 4 the difference in the magnitude for the sliding force is shown both for an 11 ft (3·35 m) deep mat and a 22 ft

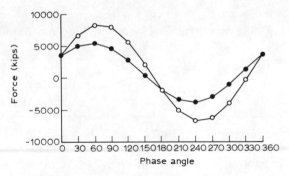

Fig. 4. Effect of mat depth on total sliding force — North Sea storm (1000 kips = 4·448 MN): ●, 11 ft mat; ○, 22 ft mat.

(6·70 m) deep mat. The greatest sliding force on the 11 ft deep mat is approximately 22·2 MN (1000 kips = 4448 MN) while for the same storm, the 22 ft deep mat has a maximum sliding force of more than 35·6 MN.

The sliding force is made up of three components. The component due to the wind is constant. The wind acts on the platform and the portion of the tower above the wave. The component due to the wave and current is varying with the particle velocity in the wave. For the North Sea storm cited and the tower in this design, the maximum value of the lateral force is approximately 10° before the wave reaches the origin or the center of the tower. The final component is made up of the hydrodynamic forces on the mat. These are a function of the particle acceleration and the volume of mass plus added mass of the mat itself. Figure 5 shows these components and Fig. 6 shows the total force as the wave passes the tower.

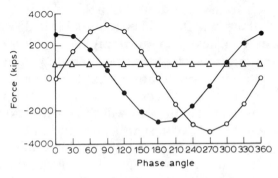

Fig. 5. Sliding force components — North Sea storm (1000 kips = 4·448 MN): ●, tower; ○, mat; △, wind.

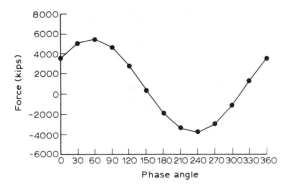

Fig. 6. Total sliding force — North Sea storm (1000 kips = 4·448 MN).

6 MAT DIMENSIONS

The next step in the preparation for the structural analysis is to determine the dimensions of the mat. For this three main factors are to be considered the first of which is the mat's resistance to overturning. There are three types of criteria for the overturning analysis. First, the simple criterion of the rig not turning over about one edge of the mat. Then there is the added requirement that there be no uplift on any edge of the mat caused by the design storm, and finally there are the reduced effective area methods such as these discussed in API RP2A.[3] These analyses are quite lengthy and not presented here.

The second criterion is a factor of safety against sliding. This has been discussed previously. A third factor is the determination of the soil types that the mat will be designed to accommodate. These three criteria will determine the footprint and depth of the mat. However, the depth could be changed at a later iteration of the design due to structural considerations.

7 MAT STRUCTURAL MODEL

The mat structure consists of bottom plate, deck plate, watertight bulkheads, and swash bulkheads. Watertight bulkheads are required for afloat conditions. Stiffeners are arranged in one direction only, with the swash bulkheads spaced to provide support. Figure 7 shows a cutaway view of the mat detailing the various plates and bulkheads. For a detail of the stiffening arrangement, Fig. 8 shows how the stiffeners are arranged.

All of the bulkheads are included in the model as membrane plates

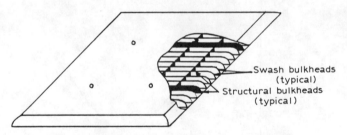

Fig. 7. Cutaway mat structural details (model).

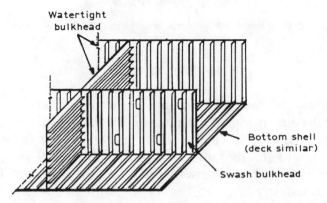

Fig. 8. Cutaway mat structural details showing stiffeners.

with orthotropic materials to account for the stiffeners being in only one direction. The actual stiffeners are not included in the model. The first bay of the tower is also modeled and connected to the mat at both the deck and the bottom plate. As described previously the loads for the various cases are applied at the top of the first bay above the mat. These loads are applied vertically as shown in Fig. 9 and horizontally as shown in Fig. 10.

The next problem is to model the soil under the mat. The soil was modeled as linear springs. However in some runs, the springs were modeled with gap elements so that if the mat lifted up, no spring force was applied to the model. The occurrence of uplift was not generally the case. The method of determining the spring constant was taken after Lloyds' method.[4] Basically, an equivalent vertical spring constant was determined for each of the types of soils listed in Table 4. The springs were attached from the ground to the bottom plate of the mat. For clay and some sand models, springs were equally spaced across the entire bottom (Fig. 11).

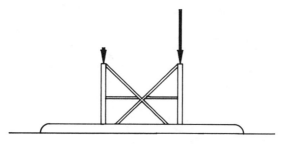

Fig. 9. Vertical loading of mat.

Fig. 10. Horizontal loading of mat through tower.

TABLE 4
Soil Types

Soft clay	2·4 kPa + 0·5 kPa gradient
Medium clay	9·6 kPa + 0·5 kPa gradient
Hard packed sand	30° angle of interval friction

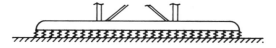

Fig. 11. Mat model support on sand and clay (100% support).

In the case of sand waves, it was assumed that the mat was supported by only 40% of the total area. The support area was equally divided, 20% on each end. In the actual analysis, the sand waves had to be modeled in both the longitudinal and transverse directions, since the mat structure did not have the same stiffness in both directions (Fig. 12).

Additional cases were run to show the effect of a mat foundation setting down over an area where an independent leg rig had been before. Figure 13 shows the footprints of a 72 m² mat superimposed on a large independent leg jackup.

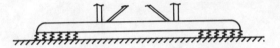

Fig. 12. Mat model support on sand waves (40% support).

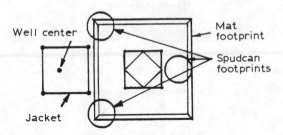

Drilling over spudcan footprints

Fig. 13. Mat placed over footprints of an independent leg jackup.

8 ANALYSIS

The method of analysis is set up on several Personal Computer based programs, spreadsheets, and databases. The finite element structural analysis is output through a modem to a database which keeps track of all elements, nodes and nodal stresses for each loading case. This database has a data sheet for each element in the model. At the moment this is restricted to the mat model and it only keeps track of the plate elements. As each load case is run, the stress data from the output of the finite element program is read by the database one element at a time. The database has six categories for stress. They are:

1. Maximum tensile stress in the X direction (any node)
2. Minimum compressive stress in the X direction (any node)
3. Maximum tensile stress in the Y direction (any node)
4. Maximum compressive stress in the Y direction (any node)
5. Maximum in-plane shear stress (any node)
6. Maximum significant stress (von Mises criteria)

A sample of the screen for one of the categories of this database is shown in Fig. 14. After the first case is input into the database, the table will be completely filled out with each category having the same information. However as subsequent loading cases are input to the database, only those categories where the desired value is exceeded are updated. This category is then completely updated with all values for the particular element and load case. Finally the load case number is updated.

Maximum Compressive Stress (× direction)

Load case	Element	Sxy	
Node	Sx	Sy	Tau

Fig. 14. Sample database category for each element.

When all of the finite element analysis load cases have been run and placed into the database, then each element will have in the database the highest stressed condition for whatever load case caused it. For example, one element may have data in each category from six different load cases. On the other hand, one load case may provide all of the data to one particular element. Also in the database the stresses are converted to an equivalent loading of the edge of the plate.

Since the finite element model has been determined by the element loading due to the overall bending of the mat, it is necessary to combine this loading with the secondary and tertiary stresses associated with the stiffened plating and the plate between the stiffeners. Thus, the next step in the analysis process utilizes a spreadsheet that calculates the allowable loading on each element accounting for these additional stresses.

Figure 15 shows how the structure reacts to the panel bottom bearing loads. This figure also shows how a typical stiffener and its associated plate is singled out for analysis, and the stresses that are calculated.[5] The stresses utilized for this case of bottom bearing pressure are as follows:

Beam end stresses
S_{1ep} — Plate stress due to bending of mat (not calculated)
S_{1es} — Stiffener stress due to bending of the mat (not calculated)
S_{2ep} — Plate stress due to bending of beam (calculated)
S_{2es} — Stiffener stress due to bending of beam (calculated)
S_{3ep} — Plate stress due to bending of plate between stiffeners (calculated)

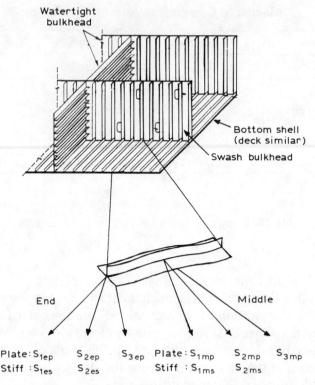

Fig. 15. Analysis of stiffener and associated plate of mat. Note that plate stresses are at the edge of the plate and the stiffener stresses are at the extreme fiber of the stiffener away from the plate.

Beam Center Stresses

S_{1mp} — Plate stress due to bending of mat (not calculated)

S_{1ms} — Stiffener stress due to bending of the mat (not calculated)

S_{2mp} — Plate stress due to bending of beam (calculated)

S_{2ms} — Stiffener stress due to bending of beam (calculated)

S_{3mp} — Plate stress due to bending of plate between stiffeners (calculated)

The next step is to calculate the allowable stresses for each element for each of the six stress categories by acceptable structural criteria. In this case, AISC[6] was utilized. For each type of stress, there is an equivalent allowable stress. Utilizing this and the fact that in storm criteria, the sum of the calculated stresses divided by the corresponding allowable stress must be less than 1·33. Thus if the end stresses on the plate were calculated, the equation would be as follows:

$$(S_{1ep}/S_{1A11}) + (S_{2ep}/S_{2A11}) + (S_{3ep}/S_{3A11}) = 1·33$$

Solving for the allowable stress due to mat bending (S_{1ep}):

$$S_{1ep} = (1.33 - (S_{2ep}/S_{2A11}) - (S_{3ep}/S_{3A11})) \times S_{1A11}$$

Similarly the three other equations can all be solved in the same manner, where:

$$S_{1mp} = (1.33 + (S_{2ep}/S_{2A11}) - (S_{3ep}/S_{3A11})) \times S_{1A11}$$
$$S_{1es} = (1.33 + (S_{2ep}/S_{2A11}) \times S_{1A11}$$
$$S_{1ms} = (1.33 - (S_{2ep}/S_{2A11}) \times S_{1A11}$$

Note that the stress due to plate bending between stiffeners always has the same sign. This is because the stress varies through the thickness of the plate and will be in tension at some point. If the deck plate were being analyzed, this plate would be in compression, since it would be the worst case. In other cases other stresses may be added to the formulae, for example the stress due to hydrostatic loads, when the vessel is afloat or is floating on the mat.

With these formulae, the spread sheet varies stiffener size, stiffener spacing and plate thickness, with a calculated bottom bearing pressure. The main purpose of the spread sheet is to calculate the maximum stress allowed in the plate due to the bending of the mat. The spreadsheet may utilize the following path.

The first step in the process would be to set up a table of plate thickness that covers the range of the finite element model. In fact, it is a good idea to carry this table slightly farther than the expected values. The second step would be to arrive at a stiffener spacing and stiffener size. This is accomplished by varying the stiffener spacing and the stiffener size in the spread sheet until the maximum allowable plate stresses are optimized. This is not as easy as one would think since in addition to the four solutions that have been presented for the stresses in the direction of the stiffeners, there are other equations that govern the stresses in the direction of the swash bulkheads. However, when this step is completed, a table of maximum allowable stresses due to mat bending is established. It should be noted that these stresses are converted to allowable loading per unit length along the element edge.

Now utilizing the data from the element database and the spreadsheet tables, it is an easy procedure to find what thickness is required for each element in each direction. By analyzing the structure in this manner it is far easier to optimize the structure and member sizes, and most importantly, this method ensures that no loading case or element will be missed in the analysis.

9 MISCELLANEOUS

There are other aspects of the structural analysis of a mat. The first and perhaps the most important detail of the entire analysis is the mat and chord interface. This is done by a separate finite element analysis, utilizing thick plate elements. This stress analysis is critical and usually dictates the fatigue life of the structure. This detail must be checked for the same soil conditions as the mat. The second aspect of any fatigue analysis is the calculation of the natural period of the structure. In this case the entire mat, tower, and platform must be modeled correctly with regard to stiffness and mass.

The soil can greatly affect the natural period of a system. For example a natural period calculation was carried out for a soft clay, a hard sand and a very hard bottom. The results showed that the natural period for the hard bottom was approximately of 5 s, for the hard sand 5·1 s and for the soft clay 7 s.

However on closer examination, the change in natural period seemed to be primarily a function of the soil rather than of the structure. A normalized deflection plot of the first mode of vibration is shown in Fig. 16. Note that there is no deflection on the hard bottom, yet on the weaker soils, there is significant deflection with a major component appearing to be rigid body rotation. A different method of modeling the soil may be necessary to arrive at a better solution. In this case, the equivalent spring stiffness of the soil dominates these calculations.

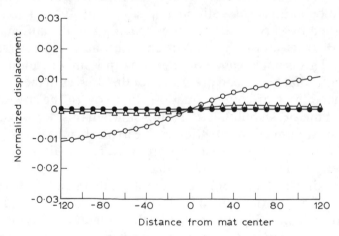

Fig. 16. Mat model shapes for various soil conditions: ●, hard bottom; ○, soft clay; △, hard sand. (1 ft = 0·3048 m.)

10 CONCLUSIONS

There is a definite need for a jackup that can operate in a multitude of soil conditions, if it is to be capable of working in all parts of the world. The method presented for analyzing the mat structure is capable of providing a good solution to the problem of analyzing various soil conditions.

REFERENCES

1. Bethlehem Steep Corporation, Technical Services Group, Environmental data on: Gulf of Mexico, North Sea and Celtic Sea, Atlantic Coast of North America, 1978.
2. Payne, M. L., Dunn, M. D., Evans, W. M., Allen, J. D., Sharples, B. P. M., Sauls, D. P. & Scales, R. E., Operator's perspective on Gulf of Mexico Jackup MODU designs. OTC Paper 5356, 1987.
3. American Petroleum Institute (API), API RP2A: Recommended practice for planning, designing, and constructing fixed offshore platforms, 18th edn, 1989.
4. Lloyd's Register of Shipping, Static and fatigue analysis for Bethlehem Sea Tower 400S drilling unit, 1986.
5. Whitley, J. O. Jr, Some aspects of the structural design of a three-column, mat supported, self elevating mobile drill platform, 1970.
6. American Institute of Steel Construction (AISC), *Manual of Steel Construction,* 8th edn, AISC, 1980.

Cyclic Storm Loading Soil–Structure Interaction for a Three Independent Leg Jack-Up

Christian Perol

CFEM Offshore Engineering, 6, Boulevard Henri Sellier, BP304, 92156 Suresnes Cedex,
France

&

Yves Meimon

Institut Francais du Petrole, 1 à 4, avenue de Bois Préau, BP311, 92506 Rueil Malmaison
Cedex, France

ABSTRACT

*A 2-D analysis of the soil–structure interaction for a jack-up platform is
presented. The special features of the analysis are:*

- *soil behaviour modelling under cyclic loading uses elasto-plastic
  equations with a multiple yield surface and kinematical hardening;*
- *the study is not executed on a single isolated spud-can, but on a more
  comprehensive structural model including the soil and a 2-D reduction
  of a complete jack-up structure;*
- *the loading history consists of some one-hundred cycles representing two
  storm sea states separated by small set-up cycles.*

*The main result is that the bottom fixity displayed during the first storm is not
significantly degraded during the second storm, even though irrecoverable
displacements and rotations occur.*

Key words: jack-up, foundations, non-linear modelling, cyclic behaviour,
storm wave.

1 INTRODUCTION

This study, although initiated three years ago is still very relevant today because bottom fixity under extreme conditions is currently being questioned by Petroleum Companies and Classification Societies. This topic is essential for improving knowledge of the behaviour of conventional jack-up platforms (not piled to the sea bed or equipped with a bottom mat) under extreme environmental conditions; a model with zero fixity gives higher bending moments in the legs at the hull connection and higher horizontal displacements at the deck level, and so will induce higher secondary effects (such as P-Delta effect).

The evaluation of bottom fixity can be done in three ways

(1) full scale measurements;
(2) centrifuge tests;
(3) numerical modelling.

The first is the most expensive and environmental conditions that should correspond to survival ones are not warranted during the recording duration.

The second cannot be performed on a complete model due to the dimensions of a real jack-up and thus only a part of the structure can be modelled, even at an acceleration of 100 g.[1] Moreover, such a study may become very difficult to carry out and interpret if a clay-like material is issued for the foundation (fabrication method, time scaling factors different for inertia forces and consolidation process).

Then, the third is the most cost effective and the most efficient, provided that all aspects of the phenomena involved can be modelled in a proper manner.

For this study, simulations as accurate as possible have been performed through a combination of extensive numerical modelling of soil, structure and loads as follows:

(1) an advanced soil model for behaviour under cyclic loading (CYCLADE model);
(2) a comprehensive structural model which is a 2-D reduction of the real structure;
(3) a realistic loading history accounting for real wave loading (cyclic loading with variable amplitude and varying from one leg to the other).

The aim of this paper is to demonstrate that such a calculation can provide useful information about the structural response and soil–structure interaction.

2 SOIL BEHAVIOUR MODELLING AND CALCULATION ALGORITHM

Under extreme environmental loading, soil element strains are irreversible, and therefore it is necessary to use a more realistic representation for the soil behaviour than the formulation based on elasticity. This was executed some nine years ago by the development of SUPER, a bidimensional computer program assuming elasto–plastic behaviour for soil elements, in which the soil–structure interface was treated by special joint elements to account for the real contact area.[2]

2.1 Cyclade soil model

SUPER was limited to monotonic loading which led Institut Francais du Petrole (IFP) to develop CYCLADE, a specific model for the behaviour of soils under non-monotonic loading. CYCLADE is an elasto-plastic model with multiple yield surfaces and kinematic hardening, a complete description of which can be found in Refs 3–5.

CYCLADE has been shown[3-4] to describe well such phenomena observed in cyclic triaxial tests as densification of sands under drained conditions or liquefaction under undrained conditions, and in 3-D true triaxial load paths.

Eleven material propery models have been incorporated in CYCLADE; three for elasticity and eight for plasticity.[5] Some of them can be directly obtained from triaxial test data, and the others (characterizing the shape of yield surfaces or the hardening state of the material) are determined by fitting experimental curves. As an objective determination of the parameters is difficult and time consuming, IFP has developed ADELAP, an automatic procedure for the determination of model parameters. Triaxial tests and general features of the material (granular shape, void ratio, water content, etc.) are stored in a data base and ADELAP first determines an initial set of parameters using general correlations. Then the final set is obtained automatically by an optimization process allowing the best reproduction of the laboratory tests. The next step consists in verifying that the parameter set is adequate for modelling tests which were not used in the ADELAP parameter identification process.[6]

2.2 FEM calculation algorithm

CYCLADE has been included in a system of IFP programs (named FONDOF) using the FEM for 2-D analysis of quasistatic behaviour of marine foundations. All the calculations are carried out incrementally, cycle-by-cycle of loading. In the presented plane strain analysis, eight-noded isoparametric elements are used and all the quantities are integrated as usual at the GAUSS points. The analysis is performed assuming infinitesimal strain theory and, as the soil response is very non-linear, a modified Newton–Raphson process is used. At each iteration, out-of-balance loads (OBL) are calculated using the rheological soil model and the boundary and loading conditions. Convergence of the process is governed by a condition on the OBL. In the case of an incremental calculation over a large number of steps, the size of the load increment must be carefully chosen in order to obtain an accurate solution.

3 OVERALL SOIL AND STRUCTURAL MODEL

The analysis is based upon a CFEM T 2005 C jack-up installed in 90 m water depth with 15 m air gap (Fig. 1) lying on a clay-like soil and submitted to two successive storm sea states representing maximum design figures for wind, current and wave in the same direction.

The spud-cans have a regular octagonal shape equivalent to a circle of 13·3 m diameter. Although the distance between two spud-cans (65·8 m) is large enough to assume their foundations do not interact, a cyclic analysis performed on an isolated spud-can cannot account for the structure behaviour. As shown in Ref. 5, such a calculation leads to unrealistic rotations ($\sim 2°$) which are not consistent with the leg and hull stiffness. Thus, a realistic study requires a model of the complete structure, and as a 3-D calculation is not possible because of the memory and CPU time requirements, the 3-D problem must be reduced to an equivalent 2-D plain strain model.

3.1 Equivalent 2-D model

The three-leg platform is reduced to two spud-cans linked together by a frame. A 3-D spud-can is replaced by an equivalent 2-D square spud-can so that the ratios of the areas and the moduli for a slice of 1 m breadth are equal. Thus, even if due to eccentric loading the spud-can does not remain in contact with the soil on its full area, these ratios remain almost identical so that the ratios of model axial and moment loads to the real structure axial and moment loads remain constant. In this study, this leads to a 11·8 m long slice (Fig. 2).

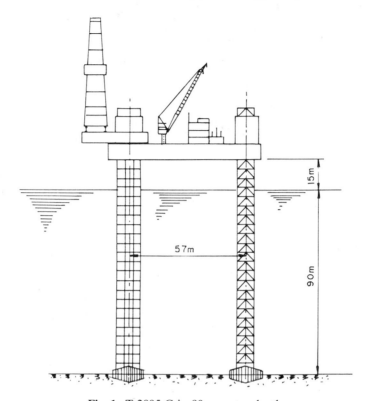

Fig. 1. T 2005 C in 90 m water depth.

The 2-D modelling of the 3-D frame is less evident. As shown in Fig. 3, two solutions seem possible:

(a) In the first model, the properties of the left leg and spud-can must be twice those of the right side. This is not realistic for the spud-can because the soil–structure contact area would not be correctly modelled. Besides, the deck inertia must vary and the loads applied to the left leg must be the combination of the loads applied to the actual left legs while the loads applied to the right leg are identical to those acting on the actual right leg. Therefore, the model is no longer symmetrical and the determination of the properties of the equivalent 1 m wide slice is not easy.

(b) The second model is obviously symmetrical and has been chosen. However, the effect of loads acting on the third leg, which are transferred through the deck, must be taken into account.

    Now the dimensions and elastic properties of the equivalent

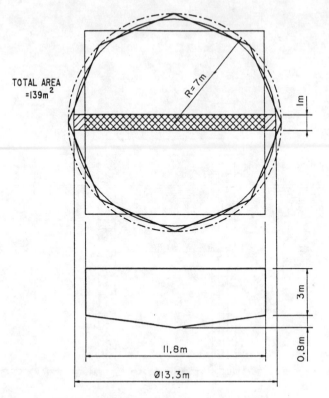

Fig. 2. 2-D reduction of the real spud-can.

frame (Fig. 4) can be deduced so that the products *ES* (where *E* is Young's modulus, *S* is the cross-sectional area) and *EI* (where *I* is the moment of inertia) are identical to those of the actual structure. To be sure that deflections and rotations of the

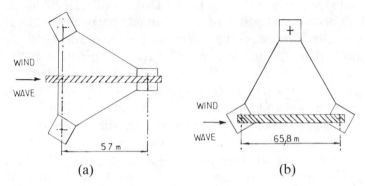

Fig. 3. Alternative 2-D slicing configurations of the real structure: (a) model 1; (b) model 2.

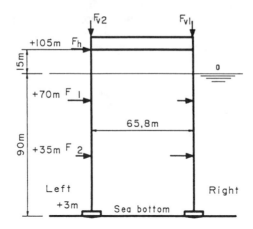

Fig. 4. Two-dimensional frame reduction of the real structure.

equivalent model will be correct, a specific calibration has been performed; a FEM elastic calculation of the equivalent structure alone under two loads cases (wave and wind) has been compared with the results of a 3-D structural model: deflections and rotations were found to be in good agreement.

3.2 Foundation material

The jack-up platform is resting on the 'Argile Noire' stiff clay which was used in a previous calculation performed on an isolated spud-can.[5] Simulations of undrained monotonic triaxial tests have been performed and have given an equivalent undrained cohesion varying almost linearly from 90 kPa at the sea bottom to 330 kPa at 50 m depth (Fig. 5). As shown in Ref. 5, CYCLADE closely reproduced undrained cyclic triaxial tests on the same material; Fig. 6 shows the good agreement obtained between the model responses and two cyclic tests conducted at the same mean deviatoric stress (qm = 60 kPa) but at two different values of the cyclic deviatoric stress (A : qc = 30 kPa, B : qc = 20 kPa).

The selected 2-D FEM mesh is plotted in Fig. 7 (it shows a deflected configuration corresponding to the maximum wave load of the first storm). It is composed of 644 nodes and 190 quadrilateral elements and, taking into account the boundary conditions (zero lateral displacement on the vertical boundaries and zero displacements at the base), the model has 1182 equations. The initial state of the soil has been calculated assuming an oedometric condition and the conic ends of the spud-cans have penetrated the soil without remolding the clay.

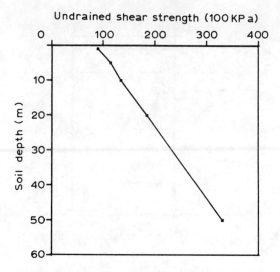

Fig. 5. Clay 'Argile Noire' shear strength profile.

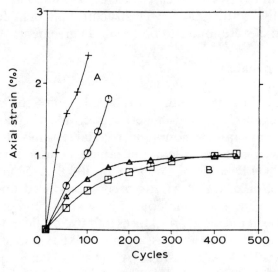

Fig. 6. Comparison between experimental and theoretical calculations of cyclic undrained triaxial tests on 'Argile Noire': experiment (+, □); calculation (○, △).

4 LOADING HISTORY

The loading history consists of the following steps:

(1) A vertical preloading of the structure to a maximum of 8820 kN (all the loads are given for a slice of 1 m breadth), applied in the drained condition and calculated in six increments.

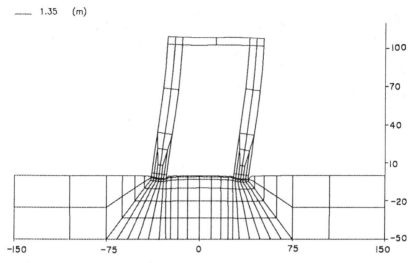

Fig. 7. Overall finite element model showing deflected configuration.

(2) Unloading to the permanent structure dead weight (6600 kN), applied in the drained condition and calculated in two increments.

(3) Undrained cyclic loading simulating two series of storm cycles separated by a series of small amplitude set up cycles. Wind and current are accounted for, assuming that they act in the same direction as the wave. The wind speeds are 45 m/s during storms and 20 m/s during the set up period.

The current velocities are 0·3 m/s at the sea bottom and 1 m/s at sea level during storms and 0·3 m/s at sea bottom and 0·5 m/s at sea level during the set up period.

As shown in Fig. 4, reduced horizontal wave loads on each leg have been applied as equivalent forces at two particular levels so as to respect both the magnitude and level of the hydrodynamic force. The reduced wind force is applied at the hull level (+105 m). The effect of the wave load on the third leg is accounted for by an additional horizontal load applied at the deck level and two opposite vertical loads applied to the legs and representing the equivalent residual overturning moment.

Normally, the duration of a typical North Sea storm is equal to 3 h which represents about 1000 cycles of loading. In order to limit computation costs, each storm has been concentrated into 53 cycles of regular wave loading as shown in Table 1. Twenty set-up cycles, corresponding to a 4 m regular wave, have been included between the two storms. Finally, Fig. 8 presents the shape of the total loading history and Fig. 9, a typical shape of the total applied moment (calculated to the origin of the reference system).

TABLE 1
Cyclic Loading Applied to the 2-D Equivalent Structure

Wave height (m)	Cycles	Steps	Lateral load (10 kN)		Moment (10 kN m)	
			Minimum	Maximum	Minimum	Maximum
8·5	15	8	5·04	21·89	1 141	2 911
11·5	7	8	5·05	28·22	1 112	3 592
15·0	3	10	−0·53	35·52	708	4 366
18·0	1	10	−5·51	46·00	595	5 469
20·0	1	11	−8·73	54·98	218	6 359
18·0	1	10	−5·51	46·00	595	5 469
15·0	3	10	−0·53	35·52	708	4 366
11·5	7	8	5·05	28·22	1 112	3 592
8·5	15	8	5·04	21·89	1 141	2 911
4·0	20	8	2·27	5·96	340	758

Each cycle has been decomposed into 8–11 steps so that the total calculation is composed of 126 cycles and 1056 steps (including the preloading–unloading steps). Each loading step required 15–30 iterations to achieve a reasonable convergence in forces (1–3%). The 1056 loading steps required 17 h CPU time on a VAX 8650 computer.

5 MAIN RESULTS

5.1 Behaviour of the soil mass

The clay mass behaviour can be characterized using isovalue plots of the shear strain (deviatoric strain), the pore pressure and the maximum plastification ratio. This last parameter is useful in identifying the zones where noticeable hardening has taken place. The main results are

(a)   the preload induces a high level of plastification around the spud-cans (higher than 90%). Then, an increase is obtained during the first storm and stabilization occurs until the end of the cyclic loading (Fig. 10). The same result is obtained for shear strain which can reach very high values (>4%) under the right edge of the spud-cans (Fig. 11). This shows the significant hardening effect of both the preloading step and the first storm. Additional storms of the same amplitude will not induce significant

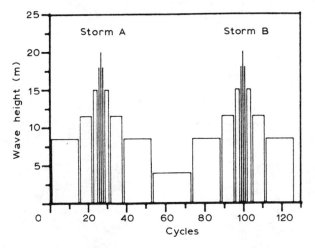

Fig. 8. Wave height profile.

degradation. It also confirms the validity of the preloading concept for installing jack-up platforms.

(b) High pore pressures are induced by the highest storm wave (up to 200 kPa) under the spud-can right edge (Fig. 12). However, at the end of the cyclic loading, the maximum pore pressure does not exceed 100 kPa and this is located in a small zone under the right spud-can.

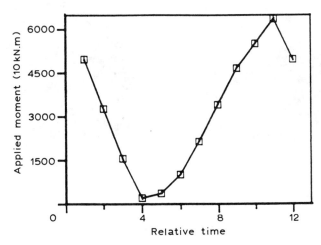

Fig. 9. Typical overall moment profile during a wave cycle.

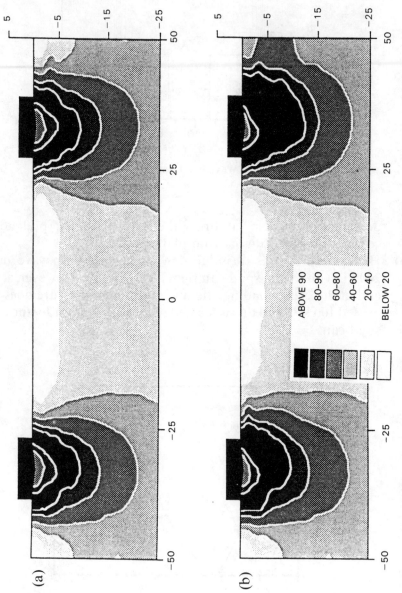

Fig. 10. Plastification ratio (%) in soil during preload (a) and during the second storm maximum wave (b).

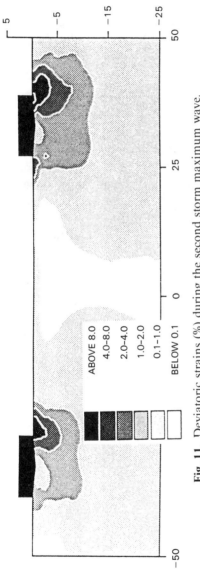

Fig. 11. Deviatoric strains (%) during the second storm maximum wave.

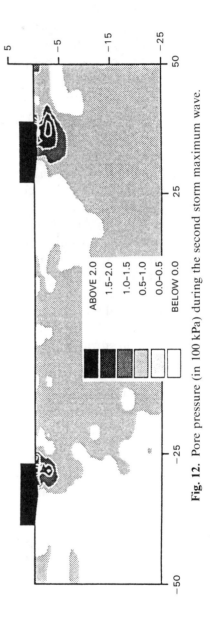

Fig. 12. Pore pressure (in 100 kPa) during the second storm maximum wave.

5.2 Behaviour of the structure

It is important to examine the evolution of the vertical load and embedment moment acting on each spud-can. This can be easily obtained by integrating the total stress just above the soil/spud-can contact line. Then, the cyclic rocking stiffness of a spud can be computed as the ratio of the embedment moment amplitude over the rotation amplitude during a cycle.

Main features of the results are

(1) The load asymmetry induces a small permanent rotation of the structure and the right spud-can is submitted to the highest vertical load (4080 kN), (Fig. 13) but less than the preload (4410 kN).

(2) Due to the hardening effect of the first storm maximum wave, stabilization of cyclic displacements is obtained for small amplitude waves (Figs 14–16). However, a further increase of displacements occurs when the second storm maximum wave is applied; a comparison with the corresponding displacements under the first storm maximum wave gives 4·9% for the lateral displacement at the hull level, 8·4% for the left spud-can rotation, and 6·4% for the right spud-can rotation. Hardening can also be qualified by the permanent displacements obtained during each storm (Table 2). An important attenuation is obtained during the second storm, especially for the rotations. As the right spud-can is

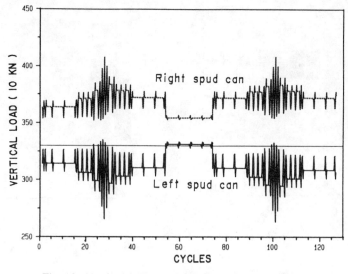

Fig. 13. Vertical loads obtained on each spud-can.

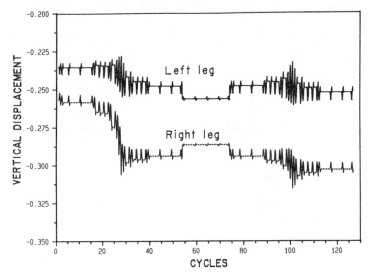

Fig. 14. Leg settlements (in m).

the most loaded, this induces greater hardening in the soil around it and, as an interesting consequence, at the end of the second storm, the relative increase of the displacements is smaller than for the left spud-can. For instance, the increment of rotation is factored by 7·4 for the right spud-can and only by 4·7 for the left.

(3) The most important result is the fact that the embedment moment of each spud-can is not significantly degraded by the second storm

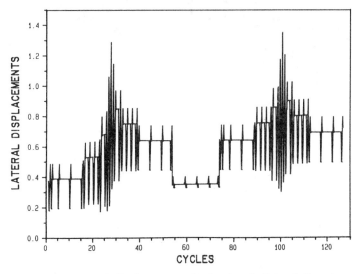

Fig. 15. Lateral displacement obtained at hull level (in m).

(a)

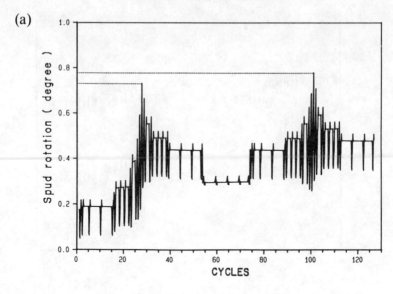

(b)

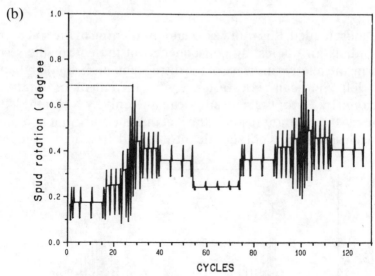

Fig. 16. Spud-can rotations: (a) right spud; (b) left spud.

loading (Figs 16 and 17); only 3·4% less for the right spud-can and 6·2% for the left one. However, as the rotation increases slightly, a small trend to decrease the rocking stiffness, Kn, is obtained (Fig. 18). The maximum decrease of Kn is reached at the maximum wave of the first storm, but is only 20%. Finally, the difference between the spud-can stiffnesses increases at the end of the loading but it does not exceed 10%. In Fig. 18, a sharp increase in

TABLE 2
Permanent Displacements Developed during each Storm Loading (Rotation in Degree)

	First storm	*Second storm*
Lateral displacement (m)	0·258	0·045
Left pile settlement (m)	0·012	0·004
Right pile settlement (m)	0·037	0·010
Left spud-can rotation	0·193	0·041
Right spud-can rotation	0·259	0·035

(a)

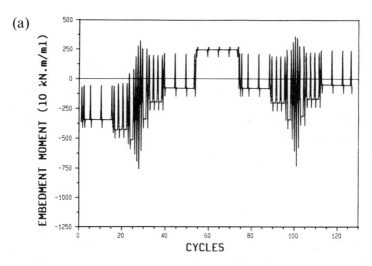

(b)

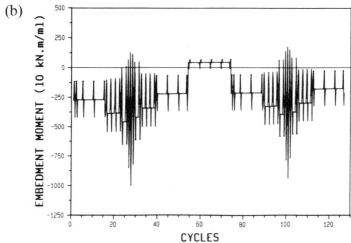

Fig. 17. Embedment moments: (a) right spud; (b) left spud.

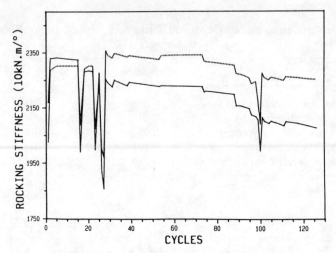

Fig. 18. Spud-can rocking stiffnesses evolution: · · · · · right spud; (b) —— left spud.

the right spud-can stiffness can be noticed just after the first storm maximum wave, it can also be understood to be a result of soil hardening as already mentioned.

6 COMPARISON WITH A SIMPLIFIED APPROACH

In the main results, the importance of the preloading concept has been emphasized. A simplified approach proposed by a Classification Society allows an evaluation of the spud-can restrainment by the sea bed as a function of the preload value. For clay, this moment is given by the following formula:

$$M/N = 0.5\,B(1 - N/N_p)$$

where

M = restraining moment
N = axial load in the leg
N_p = axial load in the leg during preloading
B = equivalent square side of the foundation

In fact, this formulation assumes that the pressure under the effective area of the foundation cannot overtake the preloading pressure (Fig. 19).

Table 3 presents a direct comparison between the FEM analysis and the above formula.

Moment 1 values are calculated directly from the above formula with

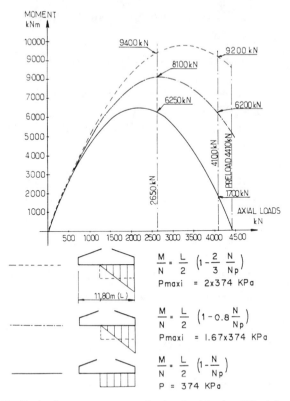

Fig. 19. Embedment moments obtained with simplified formula.

the axial load given by the FEM calculation. Moment 2 values are obtained after iterations on the axial values which vary with the restraining moments. For the right spud-can, the difference is very high.

For a closer approximation to the FEM values, the principle applied in the case of the simplified formula can be used by assuming a triangular pressure diagram under the spud-can instead of the uniform pressure diagram in such a way that the average pressure remains equal to the

TABLE 3
Comparison between FEM Analysis and Simplified Approach

Spud	Axial load (kN)	FEM moment (kN m)	Moment 1 (kN m)	Moment 2 (kN m)
Left	2 650	10 000	6 250	6 400
Right	4 100	7 500	1 700	900

preloading pressure. The simplified formula for clay would then become

$$M/N = 0.5\,B(1 - 2N/3N_p)$$

This would lead in the present case to restraining moments for both spud-cans close to 9000 kN/m. These values are closer than the previous ones but the formulation has still to be improved to avoid an overestimation for the right spud-can (+20%).

The best results are obtained using the triangular pressure diagram but with the maximum pressure limited to 1·67 times the preloading pressure (instead of twice, as above). That last assumption corresponds to the following simplified formulation:

$$M/N = 0.5\,B(1 - 0.8\,N/N_p)$$

This leads to underestimates for both values of approximately 20%.

7 CONCLUSION

An original calculation methodology of soil–structure interaction is presented. It is characterized by:

(1) modelling of the soil behaviour by an advanced constitutive equation, representing the effect of non-monotonic loadings in a proper manner;
(2) simultaneous modelling of the soil and of the structure;
(3) modelling of the loading history as accurately as possible.

Then, in the considered case of a conventional jack-up resting on stiff clay, the main conclusion is that storms do not significantly degrade either the stiffness of the foundation or the embedment moments. Also a simplified regular formula leads to underestimates of the bottom fixity especially under the most loaded leg.

Besides, irrecoverable displacements and rotations occur all through the cyclic loading which means that energy is dissipated in the soil mass. This could be accounted for in further studies by modelling inertia forces in a global dynamic analysis.

At least the importance of a realistic definition of the loading history has been demonstrated; the preload level, the number and order of wave occurrences, and the duration of the quiet phases.

REFERENCES AND BIBLIOGRAPHY

1. Lassoudiere, F. & Perol, C., Centrifuge study of soil–structure dynamic interaction for a jack-up platform submitted to sea-wave loading. Int. Conf. on Geotechnical Centrifuge Modelling. Centrifuge '88. Paris, France. 1988.
2. Meimon, Y., Thomas, P. A., Naudin, J. C. & Perol, C., Calculation of the soil structure contact of jack-up foundations. Inter. Conf. on Numerical Methods for Coupled Problems, Swansea, UK. 1981.
3. Aubry, D., Hujeux, J. C., Lassoudiere, F. & Meimon, Y., A double memory model with multiple mechanisms for cyclic behaviour of soils. Int. Symp. on Numerical Models in Geomechanics, Zurich. 1982.
4. Aubry, D., Hujeux, J. C., Lassoudiere, F. & Meimon, Y., Prediction with an elastoplastic model including multiple mechanisms for cyclic soil behaviour. Int. Workshop on Constitutive Relations for Soils. Grenoble, France. 1982.
5. Meimon, Y. & Lassoudiere, F., Application to design of marine foundations of a complete model for cyclic behaviour of offshore structures (BOSS'85). Delft, The Netherlands. 1985.
6. Meimon, Y., Lassoudiere, F. & Kodaissi, E., Fondof: a FEM software for the calculation of offshore foundations. Int. Conf. of Offshore Mechanics and Artic Eng. OMAE'87. Computer book. Houston.
7. Meimon, Y., Lassoudiere, F. & Perol, C., Modelling of the soil–structure interaction for a jack-up platform under storm loading with a cyclic elastoplastic soil model. International symposium on Modelling Soil–Water–Structure Interactions. (SOWAS 88). Delft 1988.
8. Meimon, Y. & Perol, C., Modelisation de l'interaction sol–structure pour une plateforme auto-élévatrice soumise à une tempête. Journées CLAROM 1988.

Aspects of the Stability of Jack-Up Spud-Can Foundations

G. J. M. Schotman

Koninklijke/Shell Exploratie en Produktie Laboratorium, PO Box 60, 2280 AB
Rijswijk, The Netherlands

&

M. Efthymiou*

Shell Internationale Petroleum Maatschappij BV, PO Box 162, 2501 AN The Hague,
The Netherlands

ABSTRACT

A three-level procedure for assessing jack-up foundation stability for more or less homogeneous soils is described. The objective is to provide a rational framework for these assessments that ensures their safe operation in extended year-round operations and enables their use in deeper waters than at present.

The three levels of the procedure have to be entered successively as long as foundation stability cannot be proven. The first level is a screening exercise and essentially replaces the well-known preload check. The second level compares factored foundation loads resulting from a structural analysis with foundation capacities obtained with ultimate bearing capacity formulae. The most refined third level assesses whether the displacements associated with these loads lead to an acceptable situation, i.e. capacity increase and/or load redistribution that does not result in collapse of the jack-up unit.

Since, for maximum benefit, this third-stage analysis requires a non-linear foundation model to be linked with the structural package used: such a tool is provided in the paper. Examples are given to demonstrate the impact of the assessment procedure.

This procedure forms part of the overall in-house approach to the assessment of jack-ups and has already been offered to the jack-up industry as

*Present address: Shell UK Exploration and Production, 1 Altens Farm Road, Nigg, Aberdeen AB9 2HY, UK.

part of the continuing efforts towards establishing common and accepted standards for jack-up assessments. Further developments have been identified and will be pursued.

Key words: foundation stability, site-specific assessment, spud-can foundation, load-displacement model.

1 INTRODUCTION

The procedures and criteria for the site-specific assessment of mobile jack-up units have lately been the subject of increased attention and study. The main reason for this increased interest is the diversity of the present procedures and criteria and the need to harmonise them in view of the increased use of jack-ups in deeper, more exposed waters, particularly in the North Sea and for extended year-round operations.

Attention to the foundation integrity of jack-ups has, so far, been rather patchy. In location assessments the foundation is generally considered adequate if '100% preload capacity' is available (see section 2). While this practice may be sufficient in some situations (e.g. in homogeneous sands), in many situations it is not sufficient, and hence potentially unsafe. It is felt that a more rational framework underlying the assessment of foundation stability is required.

This paper aims at providing such a framework. It discusses both the benefits and disadvantages of the often used preload check for capacity. Since this current practice often fails to define foundation integrity reliably, a three-level procedure for assessing jack-up foundation stability has been developed. The first level comprises a reformulated preload check for assessing the leeward leg, plus a sliding check for the windward leg(s). If the checks are not satisfied a second stage is entered, which involves establishing the foundation capacity to withstand combined vertical and horizontal loads, and checking whether this capacity is sufficient to safely withstand the imposed loads, appropriately factored using load and resistance factor (limit state) design principles. If the second level checks are not satisfied, a more refined third level is entered. Special attention is paid to the third level of this procedure in dealing with the inclusion of the effects of spud-can displacements on the stability. Tools are provided for taking these deflections into account. Such a displacement check has the advantage of enabling quantification of the system effects, i.e. load redistribution between the footings and hence enables the engineer to assess reserve strength rationally. Furthermore, it is a necessary tool for assessing how much moment fixity can actually be mobilised.

2 THE 100% PRELOAD CHECK AND ITS SHORTCOMINGS

The stability of a jack-up foundation is, to a very large extent, determined by the installation procedure for the unit. Usually, the unit is installed by floating it over the intended location and lowering its legs until the spud-cans contact the sea-bed. The hull is subsequently raised several metres above the sea level so that its weight is totally carried by the legs and hence by the spud-cans which penetrate the sea-bed. The preloading operation then commences, pumping water ballast into the preload tanks, thereby causing the spud-cans to penetrate further into the soil and increase their capacity. After the preloading of the foundation, the hull is elevated further to a safe level, clear of extreme wave crests, to enable the unit to operate. The water ballast used in the preloading operation is accurately known, and is one of the major characteristics for describing the capacity of the spud-can/soil system.

Hence, it is logical that the total amount of preload used in the installation procedure has, historically, often been used as a checking parameter for the spud-can's capacity to withstand extreme loads.

The 100% preload check is commonly used in the jack-up industry for the foundation assessment. It requires that the foundation reaction during preloading on any one leg should be equal to or greater than the maximum vertical reaction arising from gravity loads and 100% of the environmental loads:

$$V_{pre} > V_D + V_V + V_E \qquad (1)$$

where

V_{pre} = jack-up preload capacity per leg
V_D = vertical load reaction from fixed gravity load
V_V = vertical load reaction from maximum variable load
V_E = vertical load reaction from extreme environmental load

The key advantage of the preload check is that it is simple to use. The number of drawbacks is, however, significant. Firstly, this preload requirement assumes that the soil capacity is determined only by vertical loading, whilst in reality other load components play a role. For the case of spud-cans under leeward legs, with a high vertical load, this implies that small horizontal loads may result in failure, despite satisfaction of the preload requirement (see Fig. 1). This contradiction is even larger for footings under windward legs. Secondly, there is no safety margin to allow for

(i)   uncertainties in the evaluation of environmental loading;

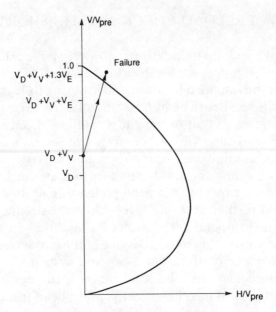

Fig. 1. The 100% preload check is potentially unsafe.

  (ii) wave heights or currents higher than those corresponding to the
        reference return period;
 (iii) possible degradation in the soil capacity under cyclic loading;
 (iv)   uneven application of the preload in the first place.

However, as will be discussed later, it is possible that a rig does not satisfy
the 100% preload requirement but, owing to stiff soil conditions, does
have an acceptable foundation stability.

To overcome these disadvantages a three-level foundation-integrity
assessment procedure has been developed, to be combined with a
location-specific structural integrity assessment.

3 INTRODUCTION TO A FOUNDATION STABILITY
ASSESSMENT PROCEDURE

3.1 Assumptions made

The procedures and criteria developed consider the footings of jack-ups
in more or less homogeneous sands and clays. Layered soil conditions,
including the case of a stronger layer overlying a weaker one (often the

cause of punch-through failures), are considered important, but have not been dealt with in this paper; neither have foundations on calcareous soils been considered specifically.

The loading condition examined is the extreme event loading, assumed to be quasi-static as far as the soil is concerned. Time-dependent effects such as consolidation, dynamic response, and creep effects are not taken into account. They are, however, qualitatively taken into account in the formulation of required partial safety factors.

3.2 Location-specific data

The first step in the foundation stability assessment procedure is the collection of site-specific data. Such information should reveal installation hazards, sea-bed stability, the presence of gas pockets, and the soil's strength and stiffness characteristics. This generally requires bathymetric surveys, side-scan sonar and/or shallow seismic surveys. Specifically, however, it requires a soil investigation programme yielding stratification and strength values for depths of at least 1·5–2·0 times the spud-can diameter below the expected penetration depth (for example, on the basis of cone penetration tests and a soil boring).

3.3 Derivation of spud-can loads

The loads to be transferred by the spud-cans from the legs to the soil follow from the structural integrity assessment of the jack-up. In such an analysis it is ordinarily assumed that the spud-cans are pinned at the base of the main spud-can structure (for example, as shown in Fig. 2). The footings are hence modelled as hinges, and the analysis results in horizontal and vertical reaction forces. Later it will be demonstrated that this model can be improved in the third level of the procedure by replacing the hinges by non-linear load/displacement models. Discussion of the extreme environmental loading falls outside the scope of this paper but may, for instance, be based on the 100-year return period.[1] Transformation to structural loads can then proceed using Stokes' fifth-order wave theory in combination with Morison's equation and appropriate drag and inertia coefficients. If the unit is dynamically sensitive, complementary dynamic analyses would need to be performed to establish the contribution from dynamics.

The foundation integrity assessment follows successively the complete or part of the three-level procedure outlined in the following section.

(a)

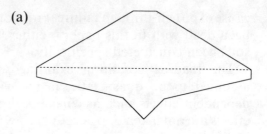

(b)

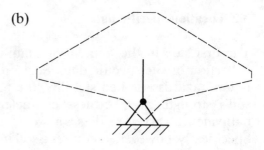

(c)

Fig. 2. Spud-cans are ordinarily modelled as hinges; only the level-III assessment allows for modelling as coupled non-linear springs; (a) typical spud-can shape; (b) modelled as hinge; (c) modelled as coupled non-linear spring (level III).

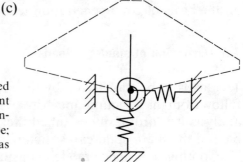

3.4 The use of partial load and resistance factors

A procedure for safeguarding against foundation failure should take into account on the one hand the loads experienced by the spud-cans and the uncertainties in these loads, and on the other the foundation resistance (capacity) and the uncertainty in this resistance. The load resistance-factor design (LRFD) approach provides a suitable framework for such a procedure. The format of the LRFD checks may be stated as

$$\frac{R_i}{\gamma_i} \geqslant \sum_{j=1}^{n} \gamma_j L_j$$

where

> R_i = characteristic resistance of component of system i
>
> γ_i = resistance factor reflecting the uncertainty in the resistance of component system i
>
> L_j = characteristic value of load type j
>
> γ_j = load factor reflecting uncertainty in load type j
>
> Σ = denotes summation over all (n) relevant load types

The assessment procedure is based on this philosophy, and hence the discussion involves not only design loads and capacities but also the associated partial factors.

4 A THREE-LEVEL ASSESSMENT PROCEDURE

It is recommended to use a three-level assessment procedure, requiring the performance of different checks in increasing order of complexity as long as the present level fails to verify foundation stability.

4.1 Level I: The preload check and the sliding check

It has been discussed that preloading the unit not only aims to achieve sufficient penetration to establish the required foundation capacity, but it also serves as a proof test and thus yields, better than any soil data, the actual capacity of the spud-can/soil system for the type of loading considered. Provided that the spud-can loading does not deviate significantly from the one experienced in the preload process, the capacity check may be replaced by a preload requirement. This is generally the case for the leeward leg spud-cans. They are exposed to large vertical loads, and for this the capacity can ultimately be proven to decrease only marginally as a result of the potential of lateral soil resistance. Even for spud-cans with shallow penetration the lateral earth pressure on the spud-can can generate about $0.1\,V_{pre}$ for typical units in clay. For sands this value is smaller, but the consequences of bearing failure are considered to be less severe because of the significant capacity increase at failure. Consequently, the foundation stability of leeward legs is proven if

$$H_E \leqslant 0.10 \times V_{pre} \tag{2}$$

$$\frac{V_{pre}}{\gamma_F} \geqslant \gamma_D V_D + \gamma_V V_V + \gamma_E V_E \tag{3}$$

where

H_E = horizontal load reaction from extreme environmental load

γ_F = partial factor covering uncertainties in the estimation of the foundation capacity on the basis of V_{pre}

$\gamma_D, \gamma_V, \gamma_E$ = partial factors covering uncertainties in fixed gravity load, variable load, and environmental load

The values of the load factors γ_D, γ_V, and γ_E, and the resistance factor γ_F are discussed at the end of this section. This preload check is shown in Fig. 3.

The preload capacity should not be used to verify the adequacy of the windward leg foundation. For these legs a sliding check has been developed. The sliding failure mode for spud-can/soil systems is of particular importance for sands. Spud-cans in clay have capacities against horizontal overloading of at least $0 \cdot 15 \, V_{pre}$, and even considerably higher when lateral earth pressure is taken into account (a total of $0 \cdot 25 \, V_{pre}$ minimum). For sands the following sliding check may be used:

$$(\gamma_D V_D + \gamma_V V_V + \gamma_E V_E) \frac{\tan \phi}{\gamma_F} \geqslant \gamma_E H_E \qquad (4)$$

where

ϕ = angle of internal friction

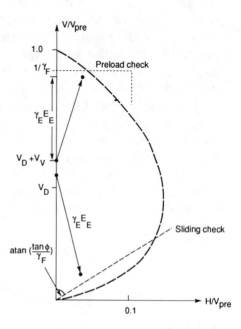

Fig. 3. Effect of the preload check and the sliding check.

γ_F = partial factor covering uncertainties in the estimation of the foundation capacity on the basis of $\tan \phi$

For the values of load and resistance factors the reader is referred to the end of this section. In view of the beneficial effect of the design variable load on stability, only half this load should be used in this check. This is reflected in the choice of $\gamma_V = 0{\cdot}5$. The sliding check is also shown in Fig. 3.

4.2 Level II: The capacity check

If the level I preload check or sliding check fails to prove integrity for the footing a more elaborate capacity check is required. The same reaction forces, resulting from the structural analysis, are compared with the bearing capacity of the spud-can foundation under combined horizontal and vertical loading. In accordance with the previously discussed format, the following is required:

$$\frac{R_F}{\gamma_F} \geqslant \gamma_D V_D + \gamma_V V_V + \gamma_E E_E \tag{5}$$

where

R_F = foundation capacity to withstand combined vertical and horizontal loads (see Fig. 4)
E_E = environmental reaction load vector (see Fig. 4), $E_E = (H_E, V_E)$
γ_F = partial factor covering uncertainties in the estimation of R_F

This capacity check essentially requires that the factored spud-can loads for all legs remain within the capacity envelope, these being factored in the direction described by the environmental load vector using γ_F. Only in the case of high-strength soil, referred to as 'stiff soil', is it sometimes possible to relax this criterion.

Most of the foundation failures associated with extreme environmental loading do not result in direct damage to the spud-can footing. When penetrating further into the soil most footings will increase their capacity (with the exception of punch-through failure), and a stable situation will be achieved. The associated displacements, however, may result in overstressing of the legs (local failure) or jack-up overturning (global failure). The failure criteria for spud-cans are thus better formulated in terms of allowable displacements than in terms of allowable loads.

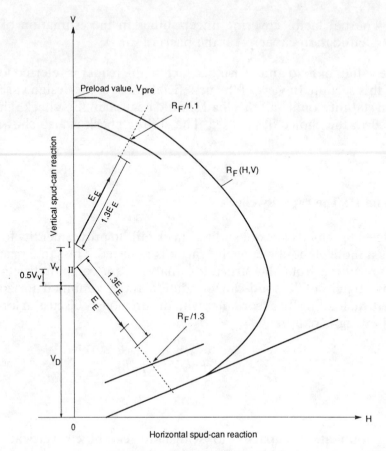

Fig. 4. The level II capacity check.

4.3 Level III: The displacement check

The third level forms the most complicated part of the foundation assessment procedure in that it requires the calculation of the displacements associated with the overload situation resulting from level II. Section 5 discusses a number of tools that may be used to perform this calculation. The resulting spud-can deflections should satisfy the following requirements.

 —Each spud-can settlement, Δv, and each spud-can horizontal displacement, Δh, should not result in an unacceptable situation with respect to resistance to overturning and strength requirements.
 —The rotation of the unit resulting from the Δv values should never exceed 2°, since such a rotation will signify the starting of a failure

process. If the operating manual imposes a more stringent limit on the allowable rotation, this more stringent limit should apply.

The loads to be used in such an analysis should be suitably factored using the same format as in levels I and II. For leeward legs the additional horizontal displacement is generally small, and the required preload resulting from the preload check (eqn (3)) may be used to assess the relevant additional spud-can settlement Δv: these checks can be based on the difference in penetration between actual preload and required preload. For windward legs the associated Δh values should ordinarily be calculated as well.

The first requirement specifically demands that it should be established whether such a load redistribution, caused by differences in spud-can displacements, can be accommodated by the structure. It is recommended, therefore that a coupled structural and non-linear foundation analysis be performed, in order to determine the effects of load redistribution adequately.

4.4 Choice of partial load and resistance factors

The partial factors used in eqns (3)–(5) are intended to cover uncertainties in the estimation of the different loads and the resistance. The load factors have been derived from LRFD considerations, and are listed in Table 1. (For the philosophy underlying these factors, see Ref. 1.)

In the choice of γ_F (the partial factor for foundation resistance) a distinction is made between the leeward and windward legs. The loading conditions for the leeward during the extreme event are similar but not identical to the situation during preloading. The major differences include

— the presence of horizontal loads during the extreme event;
— the absence of information regarding

　— horizontal loads during preload, and
　— the unequal distribution of preload over different legs;

— the inability to model time-dependent effects

　— any capacity decrease under cyclic loading,
　— effects of creep and consolidation, and
　— loading rate effects.

Owing to these uncertainties, the factor γ_F cannot be reduced to 1·0. A value of 1·1 has been considered appropriate but this value was not derived from a rigorous reliability analysis.

For the windward leg it is felt that the additional value of the preload information cannot reliably be capitalised upon using the models available at present. Thus a value of 1·3 is recommended, in line with Ref. 2.

These considerations have led to the recommended values for γ_F as given in Table 1.

Failure models and associated partial factors are the subject of continuing research. It is expected that results of this work will contribute to a further rationalisation of the foundation stability assessment procedure.

5 MODELLING TOOLS FOR THE NON-LINEAR SPUD-CAN LOAD/DISPLACEMENT BEHAVIOUR

A variety of models may be used to calculate the spud-can displacements, varying from the use of load/penetration curves to the performance of a complete non-linear analysis. This section discusses some of these models. Differences in horizontal footing displacements result in a redistribution of base shear over the different legs. Since such a redistribution usually increases the foundation capacity, a simple redistribution model will be presented as well.

5.1 Load/penetration curve check

If the preload capacity does not satisfy the inequality (Fig. 5), this implies that insufficient penetration is achieved during preloading to satisfy

TABLE 1
Partial Load and Resistance Factors

	Bearing check (governs leeward leg)		Sliding check (governs windward leg)	
	Level		Level	
	I	II	I	II
γ_D	1·0	1·0	1·0	1·0
γ_V	1·0	1·0	0·5	0·5
γ_E	1·3	1·3	1·3	1·3
γ_F	1·1	1·1	1·3	1·3
model	V_{pre}	DnV^a	$\tan\phi$	DnV^a

[a] Lateral earth pressure and side shear may be accounted for in addition to DnV.[2]

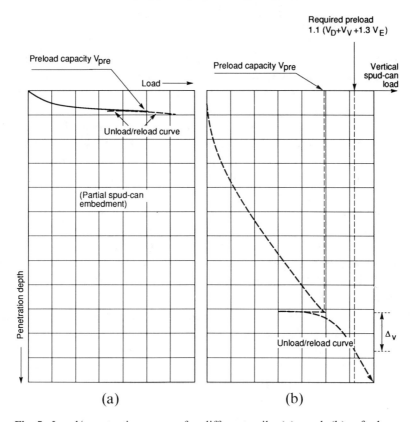

Fig. 5. Load/penetration curves for different soils: (a) sand; (b) soft clay.

foundation bearing capacity requirements. On the basis of a load/ penetration curve, calculated using a reliable ultimate bearing capacity formula, it is possible to determine the additional vertical displacement Δv required to satisfy the level I preload requirement under the leeward leg.

Figure 5 shows examples of load/penetration curves for very dense sand and soft clay. Note that, particularly in the case of soft clay, such a load/penetration curve should adequately describe the reloading response. The use of these curves follows the same philosophy as the one adopted for the preload check. It does not reflect the actual load/ penetration behaviour after reloading, which is different from those depicted in Fig. 5. However, as will be demonstrated in the next paragraph, the calculated Δv is reliable.

As demonstrated in Fig. 5(b), leeward leg penetrations may be obtained by comparing the actual preload with the required preload. This approach does not reflect the effects of the additional horizontal displacements that will occur as well.

5.2 Graphical method

This last shortcoming may be overcome by combining the use of load/
penetration curves and capacity envelopes. Although this obviously may
be done using closed form equations, a more elegant (and faster) method
is the use of graphs, as will be demonstrated.

Figure 6 includes both a load/penetration curve and an ultimate
bearing capacity curve with the vertical load V on the common vertical
axis. The inner foundation capacity curve is associated with the actual
preload capacity that is too small to satisfy the level I and II checks. The
required preload defines the second, larger curve. The spud-can's load/
displacement behaviour for the vertical direction starts to follow the
load/penetration curve. After preloading (point 1) the vertical load is
decreased to the dead and variable load only. The associated reloading
response is elastic and the penetration will consequently hardly change
(point 2). Under the extreme event environmental load vector, the
vertical load firstly increases elastically (no penetration change) until it
intersects the actual preload curve (point 3). The spud-can/soil system
now yields, and the spud-can will further penetrate the soil. Final
settlement is determined by the curve corresponding to the required
preload, because the total settlement is similar for all points on this curve.
The associated vertical load is prescribed by the intersect of load vector
and capacity curve (point 4).

This procedure illustrates the philosophy behind the load/penetration
curve check outlined in the previous section. However, it also enables the

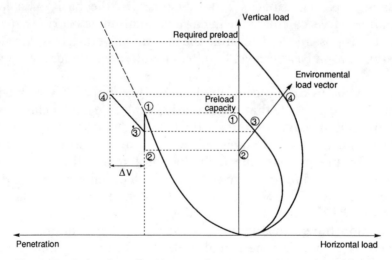

Fig. 6. Deviation from load/penetration curve under combined loads.

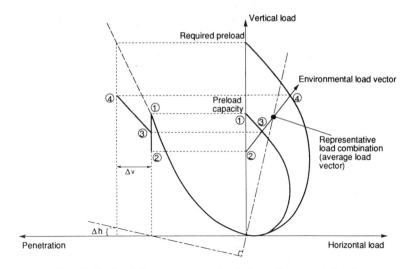

Fig. 7. Back-of-the-envelope calculation of associated Δh.

calculation of the spud-can's load/displacement characteristics (e.g. spring characteristics) in the vertical and horizontal directions. The latter may be estimated using the same graph by defining an average load vector for the load path from installation loads to extreme event (see Fig. 7). This vector roughly describes the relation between horizontal and vertical displacements, as will be discussed in the Appendix. Total horizontal displacements are thus defined by a line perpendicular to the average load vector and passing through the vertical displacement axis at the point associated with the preload settlement. Since this procedure can be applied to both windward and leeward legs, estimates of the relative displacement can be made and, consequently, upper bounds for the redistributed horizontal loads can be obtained.

5.3 Non-linear incremental load/displacement model

All the above models assume uncoupled structural and foundation stability analyses. This obviously complicates compatibility requirements. Therefore, a non-linear load/displacement model has been developed that enables a combined structural/foundation integrity assessment to be performed. The model has been calibrated with finite-element analyses and tests, and has been described extensively in Ref. 3, where it can be seen that it adequately describes key aspects of real soil behaviour without the necessity for extra soil data that are not commonly available.

The model has three degrees of freedom — the horizontal direction, the vertical direction, and rotation. The load vector $\{L\}$ and the displacement vector $\{w\}$ have been defined as

$$\{L\} = \begin{bmatrix} H \\ V \\ M \end{bmatrix} \text{ and } \{w\} = \begin{bmatrix} h \\ v \\ \theta \end{bmatrix} \tag{6}$$

They have been related to each other in the following incremental form:

$$\Delta\{L\} = K^{ep}\Delta\{w\} \tag{7}$$

In this equation, K^{ep} represents an elasto-plastic stiffness matrix which has the capacity of generating a coupled set of non-linear springs. The formulation of K^{ep} requires a description of an ultimate bearing capacity curve and knowledge of how the footing behaves at failure and after failure. The Appendix describes how K^{ep} may be constructed. It demonstrates that this model is very simple and can easily be tied in with most of the software available for performing the structural assessment of jack-ups. The format of eqn (7) has been chosen such that the stiffness matrix can easily be continued with the structural stiffness matrix. It is, however, also possible to use the inverse of the equation if required.[3]

5.4 Moment fixity

The potential impact of the model is not restricted to the acceptance of jack-ups at sites with stiff soil. It may also be used to quantify any bending moments to be transferred to the soil (note the third degree of freedom). Recent studies have increasingly indicated the existence of moment capacity under operating conditions. Since the rotational stiffness of the spud-can/soil decreases rapidly when approaching limit-state conditions, it is necessary to be able to quantify how much of this moment fixity can be mobilised under extreme events in order to be able to take it into account.

5.5 Load redistribution model

The above non-linear model has been used to validate the applicability of two analytical redistribution models that do not consider footing displacements. Figure 8 shows, for these two models, the actual preload curve and the curve associated with the required preload.

The first analytical model (Fig. 8(a)) assumes the preload curve to expand to the required preload curve for both windward and leeward legs. The first yielding spud-can foundation, usually the one at the

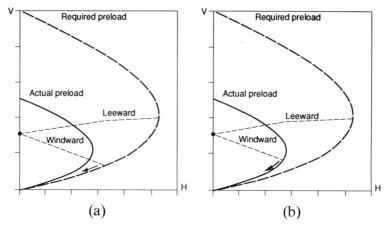

Fig. 8. Load redistribution models: (a) proposed analytical foundation model 1;
(b) proposed analytical foundation model 2.

windward leg, is taken into account by replacing the hinge in the structural model by a guided support with a horizontal reaction force dictated by the failure envelope. The vertical and horizontal reaction forces for the other supports follow from equilibrium requirements. Failure is defined by a second yielding spud-can foundation, since this results in a kinematically indeterminate model.

The second model (Fig. 8(b)) only differs from the first one in that the failure surface of the windward leg spud-can is described by the actual preload curve rather than the required preload curve.

The impact of the models has been compared with the previously described non-linear model by using them in an analysis in which a structural model was exposed to an extreme event sine wave. It is obvious that the total foundation capacity increases by allowing one spud-can to yield. Figure 9 demonstrates the effect on another important parameter, the stresses in the structure. The figure gives the calculated bending moments at the hull-leeward leg connection for an incremented environmental load factor γ_E. It is demonstrated that the second redistribution model slightly overpredicts this moment, while the first model, in many cases, leads to an underprediction. Therefore, it is recommended that the potential of load redistribution using the second model (Fig. 8(b)) be assessed.

6 EXAMPLES

The three-level stability assessment procedure has been applied to three hypothetical sites and jack-ups for demonstration purposes. For reasons

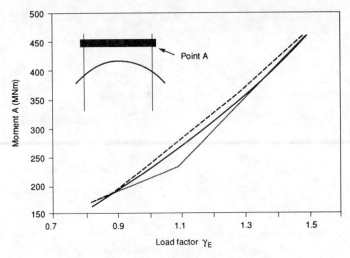

Fig. 9. Comparison of the load redistribution models with the non-linear (detailed) model: ··· analytical model 1; --- analytical model 2; —— detailed model.

of convenience, the environmental loads have been considered to be site-specific only (see Tables 2 and 3). Tables 4 and 5 contain the results of the conventional (100% preload check) and the proposed assessment procedures, respectively. The numerals in the tables indicate the level of the procedure capable of proving foundation stability, crosses show cases where foundation integrity could not be confirmed. It is concluded that both 100% preload check and the procedure cover the majority of the cases considered. There are, however, important differences: these cases (A, i), (A, iii) and (B, i), will be discussed.

The site A can be classified as a soft clay site and will generally show large spud-can penetrations. Consideration of the small jack-up unit (i) results in the conclusion that the leeward leg footings are close to limit-state conditions: the unit fails to meet the screening preload check and

TABLE 2
Characteristics of the Jack-Up Units

	Case		
	(i)	*(ii)*	*(iii)*
V_D (MN/leg)	25	30	40
V_V (MN/leg)	4	5	6
V_{pre} (MN/leg)	38	50	58
A (m²)	120	150	180

TABLE 3
Characteristics of the Three Sites

| | *Case* | | |
	(A)	*(B)*	*(C)*
c(KPa)	$10 + 0.8z$	0	200
$\phi(°)$	0	30	0
γ(kN/m^3)	8	9	8
$E_E = H_E$, V_E (MN)	(3, 6)	(4, 10)	(2, 4)

A spud-can area
c cohesion
ϕ internal friction angle
γ submerged unit weight
z depth

TABLE 4
Results of '100% Preload Check'

Case	*(i)*	*(ii)*	*(iii)*
(A)	OK	OK	OK
(B)	X	OK	OK
(C)	OK	OK	OK

OK: accepted after check
X: not accepted

TABLE 5
Results of the Proposed Assessment
Procedure

Case	*(i)*	*(ii)*	*(iii)*
(A)	X	I	II
(B)	III	III	III
(C)	I	I	I

I: accepted after level I check
II: accepted after level II check
III: accepted after level III check
X: not accepted

hence actual loads versus resistance has to be evaluated. In general, this second level will not be able to prove the integrity of units failing the preload check if considerable additional capacity (compared with the capacity following from Refs 2 and 4) is not available. This additional capacity may be available in the form of lateral earth pressure, spud-can side shear, leg/soil interaction, etc. For the soft clay considered, penetrations will range between 20 and 30 m for the units considered. For unit (i) it is not very likely that additional capacity will enable satisfaction of inequality Fig. 5. However, a more positive result may be achieved for unit (iii), for which the additional capacity available will probably result in the acceptance of the unit.

Calculation of the settlement associated with loads up to the required preload for case (A, i) results in a value of approximately 3 m for the leeward legs. Although this is a large displacement, structural checks would normally have to confirm whether or not the unit is acceptable for the site considered. This will generally yield acceptance of the units on site (B) (very dense sand), which all fail to satisfy the preload check and the capacity check but result in small additional horizontal and vertical displacements under the load path leading to the capacity envelope associated with the required preload.

Additional illustrations of the level-III assessment can be found in Ref. 3.

7 CONCLUSIONS

A three-level procedure has been developed for assessing the integrity of jack-up foundations in more or less homogeneous soils. These levels have to be entered successively as long as foundation stability cannot be proven.

Level I is essentially a screening exercise, level II the performance of a capacity check, and level III a check on footing displacements (since these eventually define the limit state of the jack-up as far as the foundation is concerned).

A number of tools have been described to enable the performance of displacement checks. In particular, a simple coupled non-linear spring model has been developed which largely facilitates the level-III check. This paper includes sufficient information to enable the use of this model. The model has an interesting spin-off in that it

— ensures complete compatibility with structural analyses;
— enables the evaluation of the foundation as a system and,

consequently, quantifies the effects of load redistribution; and
— quantifies the degree to which moment fixity is mobilised.

Consequently, it is believed that the proposed three-level procedure
could have a significant impact on the rational assessment/acceptance of
jack-up units for safe operations in deeper waters than is presently
available today.

REFERENCES

1. Leijten, S. F. & Efthymiou, M., A philosophy for the integrity assessment of
   jack-up units (SPE 19236). Paper presented at Offshore Europe Conference,
   Aberdeen, UK, September 1989.
2. Det Norske Veritas, *Rules for the design, construction and inspection of offshore
   structures*. Det Norske Veritas, Høvik, Norway, 1981.
3. Schotman, G. J. M., The effect of displacements on the stability of jack-up
   spud-can foundations (OTC 6026). Paper presented at the Offshore
   Technology Conference, Houston, USA, May 1988.
4. *Recommended Practice for Planning, Designing and Constructing Fixed Offshore
   Platforms* (17th edition). API, Washington DC, 1987.

APPENDIX

FORMULATION OF A NON-LINEAR LOAD/DISPLACEMENT MODEL

It is possible to model the non-linear load/displacement behaviour of a
footing adequately using an incremental three-degrees-of-freedom
model:

$$\{\mathbf{L}\} = \begin{bmatrix} H \\ V \\ M \end{bmatrix} \tag{A1}$$

$$\{\mathbf{w}\} = \begin{bmatrix} h \\ v \\ \theta \end{bmatrix} \tag{A2}$$

where

$\mathbf{L}$ = load vector
H = horizontal spud-can load (base shear)
V = vertical spud-can load
M = spud-can bending moment

w = displacement vector
h = horizontal displacement of spud-can
v = vertical displacement of spud-can
θ = rotation of spud-can

As described in Ref. 3 the incremental relation between {L} and {w} can be written as

$$\Delta\{L\} = K^{ep}\Delta\{w\} \tag{A3}$$

In this equation, K^{ep} is a 3×3 elasto-plastic stiffness matrix described by

$$K^{ep} = (K) - \alpha \frac{(K)\dfrac{\partial g}{\partial\{L\}} \cdot \dfrac{\partial f^T}{\partial\{L\}}(K)}{\eta + \dfrac{\partial f^T}{\partial\{L\}}(K)\dfrac{\partial g}{\partial\{L\}}} \tag{A4}$$

The equation for K^{ep} basically consists of three quantities (and their load derivatives):

— the stiffness matrix K, defining the elastic displacements;
— the plastic potential function g, describing the displacements occurring at failure;
— the yield function f, defining the limit loads on the spud-can.

The parameter η is the hardening modulus. It describes the increase in soil capacity under further spud-can penetration and can be expressed in terms of f and g:

$$\eta = -\frac{\partial f}{\partial v} \cdot \frac{\partial g}{\partial V} \tag{A5}$$

The parameter α is a switch with a value of zero for elastic behaviour and 1 in the case of plastic response (a failing spud-can). To establish a more smooth (and realistic) transition from elastic to plastic response, it can be demonstrated that other values of α may also be used (see Ref. 3).

General relationships for the stiffness matrix K, the plastic potential g, and the yield function f are shown in Fig. A1, including images of the load space and the displacement space. The following expressions for f, g, and K have been recommended.

Stiffness matrix K

This stiffness matrix is usually adequately modelled by a diagonal matrix of the form

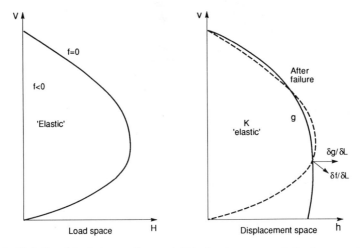

Fig. A1. Relationship between f, g, and K in load space and displacement space.

$$(K) = \begin{bmatrix} k_H & & \\ & k_V & \\ & & k_M \end{bmatrix}$$ (A6)

Expressions for the different components can, for instance, be based on Boussinesq's elastic solutions:

$$k_H = \frac{16GD(1-v)}{(7-8v)}$$ (A7)

$$k_V = \frac{2GD}{(1-v)}$$ (A8)

$$k_M = \frac{GD^3}{3(1-v)}$$ (A9)

in which D denotes the embedded spud-can diameter. The shear modulus G and Poisson's ratio v can be derived from laboratory or in-situ soil tests.

Plastic potential function *g*

The load derivatives of this function describe the failure behaviour of the spud-can, which is essentially the relation between horizontal and vertical displacements and rotations at failure. On the basis of calibration work with finite-element analyses and experiments, it has been found that the load derivatives of the function

TABLE A1
Coefficients of the Plastic Potential
Function

	Sand	Clay
ρ_1	0·50	0·40
ρ_2	$0·80\sqrt{D}$	$0·80\sqrt{D}$
V^*	$0·15\,V_{pre}$	0·0

$$g = \left(\frac{V - V^*}{V_0}\right)^2 + \left(\frac{M}{M_0} - \frac{H}{H_0}\right)^2 \qquad (A10)$$

with

$$H_0 = \rho_1 V_0$$
$$M_0 = \rho_2 V_0$$

result in realistic horizontal and vertical displacements and rotations. Recommended values for V^*, ρ_1, and ρ_2 are given in Table A1. The effect of eqn (A10) and these values is that the failure deflections are roughly proportional to the failure loads.

Yield function f

The choice of an appropriate yield function is very important but is outside the scope of this paper. Failure criteria used for the stability of shallow foundations are those included in DnV[2] or API-RP2A.[4] The ultimate bearing capacity V_{lim} resulting from these criteria has been included in the yield function as

$$f = V - V_{lim} \qquad (A11)$$

In the case of an elastic response f has a value smaller than zero, at failure f will have a value equal to zero. In the case of continued spud-can loading, f will also have to have a value of zero at the end of the load increment. In addition to ensuring that equation (A4) is obeyed, the user has to ensure that

$$f_{new} = 0 \qquad (A12)$$

These two equations define the complete spud-can load/displacement behaviour.

Switch α

In the case of modelling rigid elastic behaviour followed by plastic behaviour, the switch α has only to take values 0 and 1 for non-failure and failure, respectively. For a more gradual decrease in stiffness, as observed in small-scale tests and finite-element analyses, other values may be used. Good results have been obtained using

$$\alpha = 0 \cdot 3 - 0 \cdot 3 \cos \left[\frac{\pi(f + V_{pre}) \, \mathrm{abs}(\{L\} - \{V_{pre}\})}{V_{pre}^2} \right] f < 0 \quad \text{(A13)}$$

$$\alpha = 1 \qquad\qquad\qquad\qquad\qquad\qquad f = 0 \quad \text{(A14)}$$

This formulation reflects the stiffness decrease of the spud-can/soil system when approaching limit load conditions, and the effect of the degree by which the spud-can load deviates from the preloading load path.

Preloading of Independent Leg Units at Locations with Difficult Seabed Conditions

V. Rapoport

Marine Soil Consultants, 456 E Fair Harbor, Houston, Texas 77079, USA

&

J. Alford

Friede & Goldman, Ltd, Suite 2100, 935 Graviers Street, New Orleans, Louisiana 70112, USA

ABSTRACT

This paper, based on analyses of case histories, reviews soil conditions causing sudden uncontrolled leg penetrations, together with site data needed to plan safe installations of jack-up drilling units. Punch-through-like failure in uniform clay profile is analyzed. Preloading procedures aimed at minimizing potential damage to the unit are suggested and discussed.

Key words: foundation punch-through, spudcan, water preloading, sequential preloading, recovery procedure.

1 INTRODUCTION

In order to prevent foundation failures during storm and drilling conditions, the jack-up unit is preloaded during the initial stage of the installation procedure. During the preloading of a three leg unit, it is normally elevated to a minimal air gap (about 1·5 m) and ballast water is gradually added to the preload tanks until the weight of the unit simultaneously loads the soil under the spudcans to a level equal to or exceeding anticipated spudcan loads for the design storm condition. During the preload process, increasing soil loading usually causes the legs to penetrate deeper until the bearing capacity of the soil becomes equal to or greater than the spudcan loads. The preload process is complete when the spudcans have penetrated to a soil stratum with a bearing capacity sufficient to support the fully preloaded weight of the unit without further penetration. The preload weight is maintained for an amount of time needed to demonstrate the bearing capability of the soil; then the preload is dumped, and the unit is elevated to its working air gap. One of the main potential problems during preloading is a sudden uncontrolled leg penetration which can apply forces to the unit exceeding those imposed by the design storm.

The purpose of this paper is to review, based on the analyses of case histories, soil conditions causing sudden uncontrolled leg penetrations, and to discuss preload planning procedures which will minimize potential damage to the jack-up drilling unit.

2 SUBSURFACE SOIL CONDITIONS CAUSING SUDDEN UNCONTROLLED PENETRATIONS

There are a variety of soil conditions that could result in sudden uncontrolled penetrations. Generally, these penetrations are associated with a spudcan punching through the stronger soil layer into the underlying weaker soil (Fig. 1). This type of foundation failure could be caused by:

(a) sand, rock, or a cemented soil layer of limited thickness overlaying weaker clays, or;
(b) strong clay layer overlaying weaker clay.

Punch-through can also occur in normally consolidated or under-consolidated clays due to 'set up' caused by delays during preloading and development of an artificial crust.

During the last few years several jack-up punch-through failures were reported in mostly clay soil profiles where soil bearing capacity did not decrease with depth. Foundation analyses for some of these locations showed that the punch-through-like failures occurred in clay strata with uniform (constant) shear strength and therefore, uniform (constant) soil bearing capacity within these strata. As illustrated in Fig. 2, during preloading, after the leg load exceeds the bearing capacity of the uniform strength layer, the spudcan starts penetration into this uniform layer. This penetration of the spudcan (generally only one) causes the rig to

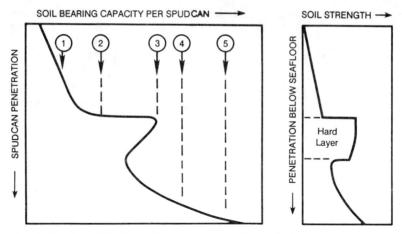

Fig. 1. Typical soil profile for sudden, uncontrolled punch-through: 1, spudcan load at partial draft; 2, spudcan load at zero draft; 3, spudcan load at punch-through; 4, spudcan load at punch-through plus tilt effect; 5, intended spudcan load at full preload.

tilt and the corresponding leg load to increase, which results in further accelerated punch-through-like penetration within the uniform bearing capacity layer until adequate soil resistance is encountered. Similar conditions can occur, as shown in Fig. 3, during storm loading following a successful preload.

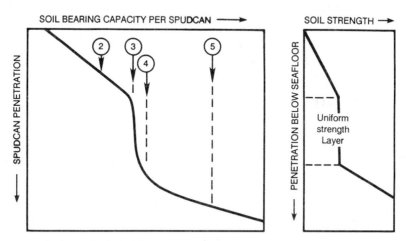

Fig. 2. Punch-through-type failure during pre-load in soils of uniform strength: 2, spudcan load at zero draft; 3, spudcan load at the beginning of penetration into the uniform strength layer; 4, spudcan load (3) plus tilt effect; 5, intended spudcan load at full preload.

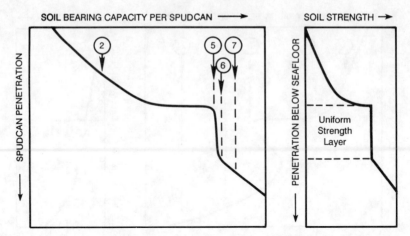

Fig. 3. Punch-through-type failure during storms in soils of uniform strength: 2, spudcan load at zero draft; 5, spudcan load at full preload; 6, spudcan load at full preload plus storm load; 7, spudcan load at full preload plus storm and tilt effects.

3 SITE INFORMATION

3.1 Subsurface information

To plan preloading operations, expected rig foundation performance should be evaluated. This evaluation requires subsurface soil data including soil stratigraphy and engineering properties. This information could be obtained from:

(a)   existing soil data for the area;
(b)   site investigation conducted from a coring vessel prior to unit arrival and;
(c)   a site investigation conducted from a jack-up unit prior to preloading which is increasingly used in industry, especially at locations with variable soil conditions.

The main results of the geotechnical site investigation should include:

(a)   estimated final spudcan penetrations;
(b)   potential problems during preloading, and for punch-through locations;
(c)   at what penetrations and leg loads punch-through could be expected;
(d)   how deep the legs will penetrate after the punch-through;
(e)   what are the criteria of the safe preloading ('. . . after x m spudcan penetration punch-through is not a problem').

3.2 Water depth information

An accurate estimate of the water depth allows one to estimate the leg penetrations and is thus a critical factor for planning and execution of preloading operations at difficult locations. It becomes especially critical at locations with high tide variations where the use of only chart datum water depth could result in highly inaccurate estimated penetrations, leading, for example, to the conclusion that the dangerous layer has been penetrated.

Water depth estimates could be made based on:

(a) existing site data (bathymetry, platform construction, etc.);
(b) measurements made during site investigations from the unit prior to preloading;
(c) reports at the location by the boats towing the jack-up, or;
(d) local tide tables offset for the actual rig location.

4 MODIFICATIONS TO STANDARD PRELOAD PROCEDURE

Because a punch-through can apply forces to a unit which exceed those imposed by the design storm, extreme care must be exercised in locations with difficult seabed conditions. Two alterations to the preload procedure have been suggested in an effort to minimize damage to a unit should this occur. The first procedure, termed 'preloading in the water' calls for adding the total preload weight while the unit is still afloat, and then slowly lifting the hull from the water. Thus, if a punch-through occurs, the buoyancy of the hull will immediately begin to relieve the forces tending to bend the legs. Since an elevating system with the capability to lift a fully preloaded hull is rarely installed on jack-up drilling units, this procedure is not applicable to the preload of most units. However, in locations with reasonable tidal variation, a falling tide can be used in a similar manner to apply the preload weight to the unit. In this procedure, the full preload weight is added to the hull while it is still in the water. As the legs penetrate, the hull is raised to maintain the selected draft. When full preload is onboard, the hull is elevated to the least draft allowed by the capacity jacking system. As the tide rises, the hull is elevated maintaining this least draft. As the tide falls, jacking is suspended and the falling tide gradually applies the preload to the legs.

A second procedure, termed 'sequential preloading', calls for preloading only one leg at a time. After the preload procedure is completed for the first leg, it is then repeated for the second and then the third leg.

TABLE 1
Punch-through Study

Hull angle	Punch-through	Wgt on fwd leg	Moment on fwd leg	Buoyancy force	Moment on aft leg	Wgt on aft leg	Hull draft	Weight lever	Fwd leg lever	Buoyancy force	Buoyancy lever	P-δ	Assumed δ
0·00	0·00	8 437	0	0	0	8 438	−0·00	38·33	115·00	0·00	0·00	0·000	0·000
0·20	0·40	8 312	3 346	419	3 338	8 291	0·00	38·82	115·00	419·30	64·44	0·019	0·019
0·40	0·80	8 274	6 660	620	6 607	8 209	0·00	39·30	115·00	620·20	70·53	0·037	0·037
0·60	1·20	8 231	9 935	822	9 812	8 130	0·01	39·79	114·99	821·72	74·19	0·055	0·055
0·80	1·61	8 184	13 164	1 031	12 947	8 049	0·01	40·27	114·99	1 030·56	76·23	0·073	0·073
1·00	2·01	8 133	16 346	1 250	16 008	7 965	0·02	40·76	114·98	1 249·99	77·04	0·090	0·090
1·20	2·41	8 087	19 496	1 459	19 005	7 883	0·03	41·24	114·97	1 459·05	77·66	0·107	0·107
1·40	2·81	8 018	22 542	1 697	21 924	7 798	0·04	41·72	114·97	1 697·16	78·20	0·124	0·124
1·60	3·21	7 957	25 553	1 924	24 778	7 716	0·05	42·21	114·96	1 924·36	78·51	0·140	0·140
1·80	3·61	7 894	28 505	2 154	27 563	7 633	0·06	42·69	114·94	2 153·64	78·72	0·156	0·156
2·00	4·01	7 826	31 385	2 385	30 283	7 551	0·07	43·17	114·93	2 384·93	78·93	0·172	0·172
2·20	4·41	7 756	34 200	2 618	32 934	7 469	0·09	43·65	114·92	2 618·15	79·06	0·187	0·187
2·40	4·82	7 683	36 937	2 853	35 522	7 338	0·11	44·13	114·90	2 853·25	79·21	0·202	0·202
2·60	5·22	7 611	39 622	3 090	38 034	7 306	0·12	44·61	114·88	3 090·03	79·15	0·216	0·216
2·80	5·62	7 530	42 193	3 329	40 498	7 227	0·14	45·09	114·86	3 328·54	79·33	0·230	0·230
3·00	6·02	7 450	44 702	3 569	42 888	7 147	0·16	45·57	114·84	3 568·77	79·35	0·244	0·244
3·20	6·42	7 366	47 123	3 811	45 217	7 068	0·19	46·05	114·82	3 810·70	79·39	0·257	0·257
3·40	6·82	7 278	48 445	4 054	47 487	6 990	0·21	46·53	114·80	4 054·45	79·45	0·270	0·270
3·60	7·22	7 190	51 690	4 300	49 690	6 911	0·24	47·01	114·77	4 299·85	79·45	0·283	0·283
3·80	7·62	7 099	53 841	4 547	51 831	6 834	0·26	47·48	114·75	4 546·72	79·43	0·295	0·295
4·00	8·02	7 005	55 897	4 795	53 911	6 756	0·29	47·96	114·72	4 795·08	79·42	0·307	0·307
4·20	8·42	6 909	57 852	5 045	55 932	6 679	0·32	48·44	114·69	5 044·90	79·40	0·318	0·318
4·40	8·82	6 809	59 695	5 298	57 891	6 603	0·35	49·91	114·66	5 297·95	79·37	0·329	0·329

4·60	9·22	6 708	61 455	5 550	6 527	59 793	0·38	49·39	114·63	5 549·96	79·33	0·340	0·340
4·80	9·62	6 600	63 051	5 806	6 453	61 651	0·42	49·86	114·60	5 806·26	79·36	0·350	0·350
5·00	10·02	6 493	64 572	6 063	6 378	63 437	0·45	50·33	114·56	6 063·13	79·31	0·360	0·360
5·20	10·42	6 382	65 977	6 321	6 304	65 169	0·49	50·81	114·53	6 321·45	79·27	0·369	0·369
5·40	10·82	6 270	67 268	6 581	6 231	66 844	0·53	51·28	114·49	6 581·24	79·22	0·378	0·378
5·60	11·22	6 154	68 421	6 843	6 158	68 467	0·57	51·75	114·45	6 843·49	79·18	0·387	0·387
5·80	11·62	6 032	69 427	7 107	6 086	70 047	0·61	52·22	114·41	7 107·23	79·17	0·395	0·395
6·00	12·02	5 911	70 333	7 371	6 015	71 567	0·65	52·69	114·37	7 371·36	79·13	0·402	0·402
6·20	12·42	5 787	71 101	7 638	5 944	73 033	0·69	53·16	114·33	7 637·97	79·09	0·409	0·409
6·40	12·82	5 659	71 727	7 908	5 873	74 441	0·74	53·63	114·28	7 907·71	79·03	0·416	0·416
6·60	13·22	5 530	72 239	8 177	5 803	75 802	0·78	54·10	114·24	8 176·54	78·97	0·422	0·422
6·80	13·62	5 395	72 558	8 448	5 735	77 129	0·83	54·57	114·19	8 448·49	78·95	0·428	0·428
7·00	14·01	5 258	72 752	8 722	5 666	78 399	0·88	55·03	114·14	8 721·88	78·90	0·434	0·434
7·20	14·41	5 119	72 800	8 997	5 598	79 615	0·93	55·50	114·09	8 997·36	78·85	0·439	0·439
7·40	14·81	4 978	72 711	9 273	5 531	80 788	0·98	55·97	114·04	9 272·99	78·80	0·443	0·443
7·60	15·21	4 831	72 427	9 552	5 465	81 923	1·03	56·43	113·99	9 551·86	78·76	0·447	0·447
7·80	15·61	4 684	72 020	9 831	5 399	83 007	1·09	56·89	113·94	9 830·83	78·71	0·451	0·451
8·00	16·00	4 533	71 435	10 112	5 333	84 047	1·14	57·36	113·88	10 112·40	78·66	0·454	0·454
8·20	16·40	4 377	70 653	10 396	5 269	85 053	1·20	57·82	113·82	10 396·39	78·62	0·456	0·456
8·40	16·80	4 221	69 736	10 681	5 206	86 009	1·26	58·28	113·77	10 680·81	78·56	0·458	0·458
8·60	17·20	4 061	68 644	10 967	5 142	86 928	1·32	58·74	113·71	10 966·66	78·51	0·460	0·460
8·80	17·59	3 899	67 390	11 255	5 079	87 793	1·38	59·21	113·65	11 254·92	78·45	0·461	0·461
9·00	17·99	3 729	65 864	11 546	5 019	88 656	1·44	59·67	113·58	11 545·82	78·43	0·462	0·462

Depth + penetration = 110·00, Hull WGT = 9509·40, VarLd WGT 2600·00, PreLd WGT 9789·47, Leg WGT = 3413·60, Tot WGT = 25 312·47.
Air gap = 0·00, Hull LCG = 106·67, VarLd LCG 106·67, PreLd LCG 106·67, Leg LCG = 106·67, Tot LCG = 106·67.
Units = English, Hull VCG = 24·28, VarLd VCG 25·00, PreLd VCG 13·00, Leg VCG = 94·06, Tot VCG = 29·40.
Angle increment = 0·20, Angle = 15·37.

The theory behind this procedure is that since it is the weight of the unit that effects damage to the legs, the damaging forces will be less if only one third of the preload weight is onboard at any one time. Unfortunately, the sequential action of this procedure can increase the time required to preload by 100–200%.

In order to compare the relative merits of these two procedures, a spreadsheet program for a personal computer was used to compute forces and moments acting on the legs of a Friede & Goldman L-780 MOD II jack-up unit for various combinations of water depth, air gap, preload weight, and hull angle. The program considered the effect of hull buoyancy (which was input in tabular form from a stability program) and leg/hull deflections, the P-δ effect. A typical program spreadsheet is

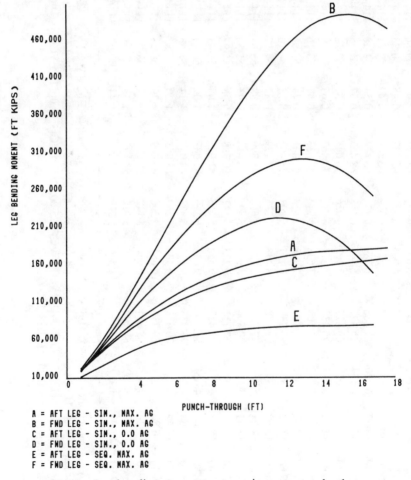

A = AFT LEG - SIM., MAX. AG
B = FWD LEG - SIM., MAX. AG
C = AFT LEG - SIM., 0.0 AG
D = FWD LEG - SIM., 0.0 AG
E = AFT LEG - SEQ. MAX. AG
F = FWD LEG - SEQ. MAX. AG

Fig. 4a. Leg bending moments at maximum water depth.

presented in Table 1. A graphical comparison of the effectiveness of these two procedures in reducing punch-through leg bending moments at maximum water depth is shown in Figs 4a and 4b. The program indicates that the estimated penetration before damage is increased by about 42% by changing the preload procedure from simultaneous preload at maximum air gap to simultaneous preload at 0·0 air gap. It also indicates that a change from simultaneous preload at maximum air gap to sequential preload at maximum air gap will increase the estimated punch-through by about 14%. However, application of both procedures, that is sequential preload at 0·0 air gap, reduces leg loads to allowable values for the entire range of punch-through investigated. This clearly indicates that there is substantial benefit to altering the preload

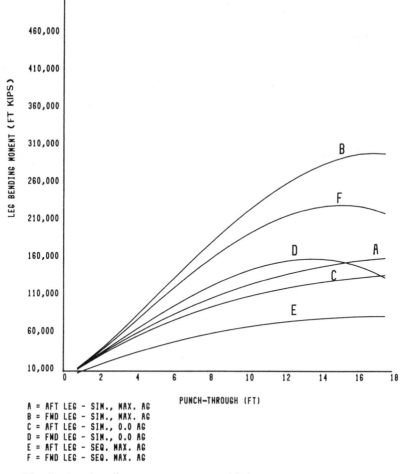

A = AFT LEG - SIM., MAX. AG
B = FWD LEG - SIM., MAX. AG
C = AFT LEG - SIM., 0.0 AG
D = FWD LEG - SIM., 0.0 AG
E = AFT LEG - SEQ. MAX. AG
F = FWD LEG - SEQ. MAX. AG

Fig. 4b. Leg bending moments at two-thirds maximum water depth.

procedure at deep water locations with difficult seabed conditions.

A third procedure which has been suggested is to preload sequentially with the preloading leg raised somewhat higher than the other two legs. The reasoning behind this procedure is that because the preloading leg is higher, it can penetrate further before the critical bending forces are reached. Unfortunately, this procedure may worsen the punch-through damage if it is misapplied. The jacking system guides will allow only a fixed amount of angularity between the legs and the hull (about $0.1°$ on the L-780 MOD II units); if this limit is exceeded, a bending moment is induced in the preloading leg which is additional to the leg bending moment imposed by a punch-through.

5 PUNCH-THROUGH RECOVERY PROCEDURE

These procedures will minimize the damage caused by a punch-through, but none of these will prevent a punch-through from occurring. If there is a punch-through, preventing damage during the re-leveling of the unit is as important as preventing damage during the punch-through. To understand how damage can occur to units with fixed jacking systems, that is units whose jacking systems are fimly welded to the hull, one must first understand the leg guide system and the characteristics of the electric induction motors which power the jacking system. In a punch-through, the leg loads no longer act along the axis of the legs (see Fig. 5).

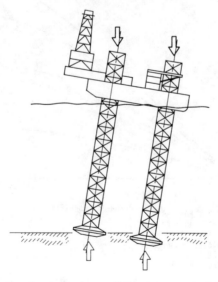

Fig. 5. Load misalignment following punch-through.

This out of level gravity loading causes the jacking system pinions to be loaded unequally. The load on the high side pinions is increased, and the load on the low side pinions is decreased. This unequal pinion loading is a moment which balances the leg bending moment caused by the non-axial component of leg loading. This pinion loading acts vertically along the strength axis of the leg, and this is the most desirable way of reacting a leg bending moment.

There is, however, another load path for balancing a leg bending moment, and that is the leg guide system. However the leg guides do not apply loads along the strength axis of the legs; they apply loads horizontally along the weak leg axis.

The proportionate share of the loads going to the pinions and the guides depends on the relative stiffness of each system, and calculations show that the L-780 MOD II pinions take about 90% of the bending moment and the guides about 10% of the bending moment.

All of this assumes that the jacking system motors have not been running with a bending moment acting on the leg. If the motors are operated, the leg loading situation is radically altered. This is because of the speed–load characteristics of the jacking system induction motors (see Fig. 6). The speed of the motors is dependent on their load; if the motors have different loads, they will run at different speeds. A view of the loads on the leg of an inclined hull is presented in Fig. 7.

If the hull is lowered in this condition, the motors on the high side which have higher load will run faster than the motors on the low side. Because one side of the leg is moving faster than the other, the leg will rotate with respect to the hull causing it to press harder on the guides. This load transfer will continue until all of the bending moment has been transferred to the guides and all pinions are equally loaded. Unfortunately, this means that the bending moment has been transferred from the strength axis of the leg to the weak axis of the leg.

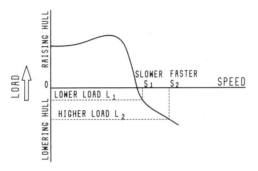

Fig. 6. Speed–load characteristics of induction motors.

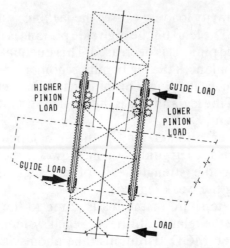

Fig. 7. Leg loads for an inclined hull.

How then can the hull be re-leveled if the jacking system cannot be used? The hull can be leveled by altering the jacking system so that the pinions will maintain unequal loading during movement. This can be done on the L-780 MOD II by disconnecting the motors on the tension chords (the chords facing the direction of the punch-through leg), but keeping the brakes electrically active. Thus, when the hull is jacked down, the tension chord motors will have no load and the total weight of the rig will be supported by the compression chord motors. This unequal pinion loading causes a moment which will tend to react the leg bending moment along the strength axis of the leg. When the hull is within 4° of level, the bending moment is reduced to a value which allows the hull to be leveled with all motors active.

6 SUMMARY

Successful preloading of a jack-up unit at locations with difficult seabed conditions requires careful planning. A thorough knowledge of the site data is required; this includes water depth, tidal variations and soil bearing capacity. If a punch-through potential is indicated, two procedures, 'preloading in the water' and sequential preloading' may be employed to minimize damage to the unit. If a punch-through does occur, selective use of the jacking system motors will aid in minimizing damage to the unit during re-leveling.

Index